HEINEMANN MODULAR MATHEMATICS
for EDEXCEL AS AND A-LEVEL

Revise for Core Mathematics 2

Greg Attwood   Alistair Macpherson   Bronwen Moran
Joe Petran   Keith Pledger   Geoff Staley   Dave Wilkins

heinemann.co.uk
✓ Free online support
✓ Useful weblinks
✓ 24 hour online ordering

01865 888058

Heinemann Educational Publishers
Halley Court, Jordan Hill, Oxford OX2 8EJ
Part of Harcourt Education

Heinemann is the registered trademark of
Harcourt Education Limited

© Greg Attwood, Alistair David Macpherson, Bronwen Moran, Joe Petran, Keith Pledger, Geoff Staley,
Dave Wilkins 2005

First published 2005

09 08 07 06
10 9 8 7 6 5 4 3 2

British Library Cataloguing in Publication Data is available
from the British Library on request.

10-digit ISBN: 0 435511 23 8
13-digit ISBN: 978 0 435511 23 4

Designed by Bridge Creative Services
Typeset by Tech-Set Ltd

Original illustrations © Harcourt Education Limited, 2005

Illustrated by Tech-Set Ltd

Cover design by Bridge Creative Services

Printed in China by China Translation & Printing Services Ltd.

Acknowledgements
Every effort has been made to contact copyright holders of material reproduced in this book. Any omissions will
be rectified in subsequent printings if notice is given to the publishers.

About this book

This book is designed to help you get your best possible grade in your Core 2 examination. The authors are Chief and Principal examiners, and have a good understanding of Edexcel's requirements.

Revise for Core 2 covers the key topics that are tested in the Core 2 exam paper. You can use this book to help you revise at the end of your course, or you can use it throughout your course alongside the course textbook, *Heinemann Modular Mathematics for Edexcel AS and A-level Core 2*, which provides complete coverage of the syllabus.

Helping you prepare for your exam

To help you prepare, each topic offers you:

- **Key points to remember** – summarise the mathematical ideas you need to know and be able to use.

- **Worked examples and examination questions** – help you understand and remember important methods, and show you how to set out your answers clearly.

- **Revision exercises** – help you practise using these important methods to solve problems. Exam-level questions are included so you can be sure you are reaching the right standard, and answers are given at the back of the book so you can assess your progress.

- **Test Yourself questions** – help you see where you need extra revision and practice. If you do need extra help, they show you where to look in the *Heinemann Modular Mathematics for Edexcel AS and A-level Core 2* textbook and which example to refer to in this book.

Exam practice and advice on revising

Examination style paper – this paper at the end of the book provides a set of questions of examination standard. It gives you an opportunity to practise taking a complete exam before you meet the real thing. The answers are given at the back of the book.

How to revise – for advice on revising before the exam, read the How to revise section on the next page.

How to revise using this book

Making the best use of your revision time

The topics in this book have been arranged in a logical sequence so you can work your way through them from beginning to end. But **how** you work on them depends on how much time there is between now and your examination.

If you have plenty of time before the exam then you can **work through each topic in turn**, covering the key points and worked examples before doing the revision exercises and test yourself questions.

If you are short of time then you can **work through the Test Yourself sections** first, to help you see which topics you need to do further work on.

However much time you have to revise, make sure you break your revision into short blocks of about 40 minutes, separated by five- or ten-minute breaks. Nobody can study effectively for hours without a break.

Using the Test Yourself sections

Each Test Yourself section provides a set of key questions. Try each question:

- If you can do it and get the correct answer, then move on to the next topic. Come back to this topic later to consolidate your knowledge and understanding by working through the key points, worked examples and revision exercises.

- If you cannot do the question, or get an incorrect answer or part answer, then work through the key points, worked examples and revision exercises before trying the Test Yourself questions again. If you need more help, the cross-references beside each Test Yourself question show you where to find relevant information in the *Heinemann Modular Mathematics for Edexcel AS and A-level Core 2* textbook and which example in *Revise for C2* to refer to.

Reviewing the key points

Most of the key points are straightforward ideas that you can learn: try to understand each one. Imagine explaining each idea to a friend in your own words, and say it out loud as you do so. This is a better way of making the ideas stick than just reading them silently from the page.

As you work through the book, remember to go back over key points from earlier topics at least once a week. This will help you to remember them in the exam.

Algebra and functions

Key points to remember

1 If f(x) is a polynomial and f(a) = 0, then ($x - a$) is a factor of f(x).

2 If f(x) is a polynomial and $f\left(\dfrac{b}{a}\right) = 0$, then ($ax - b$) is a factor of f($x$).

3 If a polynomial f(x) is divided by ($ax - b$) then the remainder is $f\left(\dfrac{b}{a}\right)$.

Example 1

Simplify $\dfrac{x^2 + 2x - 3}{2x^2 + 3x - 9}$.

Factorise $x^2 + 2x - 3$

$$x^2 + 2x - 3 = x^2 + 3x - x - 3$$
$$= x(x + 3) - 1(x + 3)$$
$$= (x + 3)(x - 1)$$

> $ac = -3$ and $3 + (-1) = 2\,[= b]$
> Factorise
> Common factor is ($x + 3$)

Factorise $2x^2 + 3x - 9$

$$2x^2 + 3x - 9 = 2x^2 + 6x - 3x - 9$$
$$= 2x(x + 3) - 3(x + 3)$$
$$= (2x - 3)(x + 3)$$

> $ac = -18$ and $6 + (-3) = 3\,[= b]$
> Factorise
> Common factor is ($x + 3$)

$$\frac{x^2 + 2x - 3}{2x^2 + 3x - 9} = \frac{(x + 3)(x - 1)}{(2x - 3)(x + 3)}$$

$$= \frac{(x + 3)(x - 1)}{(2x - 3)(x + 3)}$$

> Divide top and bottom by ($x + 3$)

$$= \frac{x - 1}{2x - 3}$$

Example 2

Divide $x^3 + 5x^2 + 4x - 4$ by $(x + 2)$.

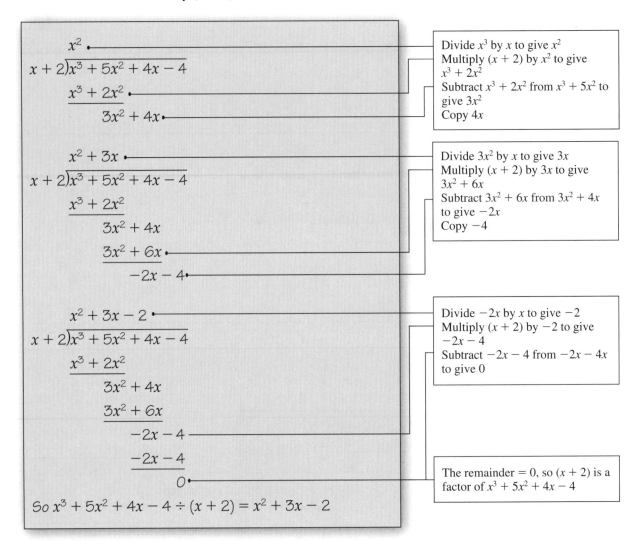

$$\text{So } x^3 + 5x^2 + 4x - 4 \div (x + 2) = x^2 + 3x - 2$$

Example 3

Show that $(x - 3)$ is a factor of $x^4 - 10x^2 + 9$. Hence express $x^4 - 10x^2 + 9$ in the form $(x - 3)(x^3 + Ax^2 + Bx + C)$, where the values of A, B and C are to be found.

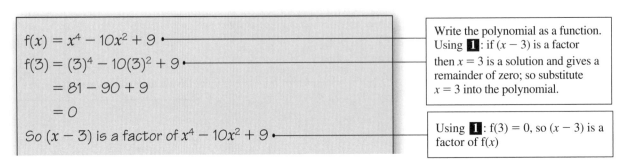

$$\begin{array}{r} x^3 + 3x^2 - x - 3 \\ x - 3 \overline{\smash{\big)}\, x^4 + 0x^3 - 10x^2 + 0x + 9} \end{array}$$

Divide $x^4 - 10x^2 + 9$ by $(x - 3)$. Remember to use $0x^3$ and $0x$ so that the sum is laid out correctly.

$$\underline{x^4 - 3x^3}$$
$$3x^3 - 10x^2$$
$$\underline{3x^3 - 9x^2}$$
$$-x^2 + 0x$$
$$\underline{-x^2 + 3x}$$
$$-3x + 9$$
$$\underline{-3x + 9}$$
$$0$$

Remember you can check your division here. As $(x - 3)$ is a factor of $x^4 - 10x^2 + 9$ the remainder must be zero.

So $x^4 - 10x^2 + 9 = (x - 3)(x^3 + 3x^2 - x - 3)$

Notice $A = 3$, $B = -1$ and $C = -3$

Example 4

Factorise $3x^3 + 5x^2 - 26x + 8$ completely.

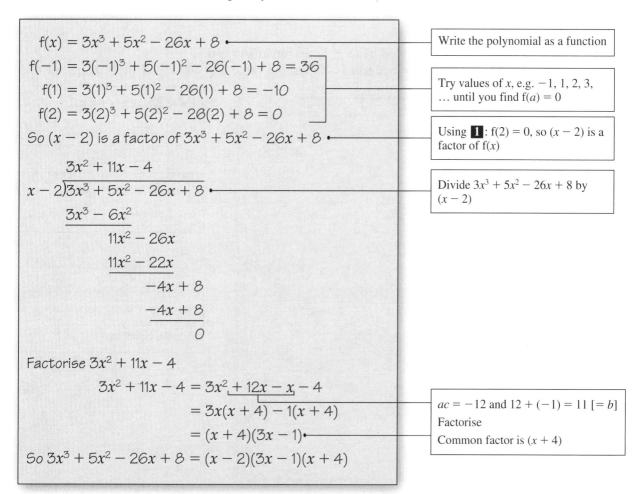

$f(x) = 3x^3 + 5x^2 - 26x + 8$

Write the polynomial as a function

$f(-1) = 3(-1)^3 + 5(-1)^2 - 26(-1) + 8 = 36$
$f(1) = 3(1)^3 + 5(1)^2 - 26(1) + 8 = -10$
$f(2) = 3(2)^3 + 5(2)^2 - 26(2) + 8 = 0$

Try values of x, e.g. -1, 1, 2, 3, … until you find $f(a) = 0$

So $(x - 2)$ is a factor of $3x^3 + 5x^2 - 26x + 8$

Using **1** : $f(2) = 0$, so $(x - 2)$ is a factor of $f(x)$

$$\begin{array}{r} 3x^2 + 11x - 4 \\ x - 2 \overline{\smash{\big)}\, 3x^3 + 5x^2 - 26x + 8} \end{array}$$

Divide $3x^3 + 5x^2 - 26x + 8$ by $(x - 2)$

$$\underline{3x^3 - 6x^2}$$
$$11x^2 - 26x$$
$$\underline{11x^2 - 22x}$$
$$-4x + 8$$
$$\underline{-4x + 8}$$
$$0$$

Factorise $3x^2 + 11x - 4$

$3x^2 + 11x - 4 = 3x^2 + 12x - x - 4$

$ac = -12$ and $12 + (-1) = 11\ [= b]$
Factorise

$= 3x(x + 4) - 1(x + 4)$

$= (x + 4)(3x - 1)$

Common factor is $(x + 4)$

So $3x^3 + 5x^2 - 26x + 8 = (x - 2)(3x - 1)(x + 4)$

Worked exam style question 1

When $8x^3 + ax^2 + 5$ is divided by $(2x + 1)$ the remainder is 2. Find the value of a.

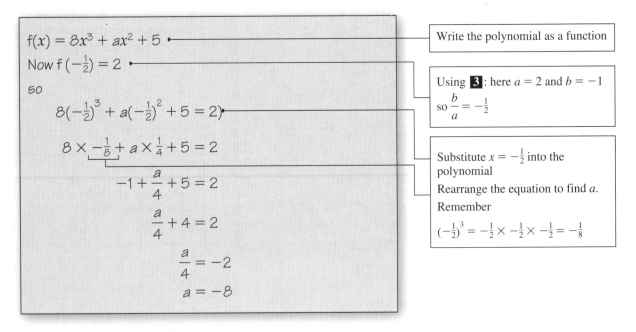

$f(x) = 8x^3 + ax^2 + 5$ ── Write the polynomial as a function

Now $f\left(-\frac{1}{2}\right) = 2$ ── Using **3**: here $a = 2$ and $b = -1$ so $\frac{b}{a} = -\frac{1}{2}$

so

$8\left(-\frac{1}{2}\right)^3 + a\left(-\frac{1}{2}\right)^2 + 5 = 2$

$8 \times -\frac{1}{8} + a \times \frac{1}{4} + 5 = 2$

$-1 + \frac{a}{4} + 5 = 2$

$\frac{a}{4} + 4 = 2$

$\frac{a}{4} = -2$

$a = -8$

Substitute $x = -\frac{1}{2}$ into the polynomial

Rearrange the equation to find a.

Remember

$\left(-\frac{1}{2}\right)^3 = -\frac{1}{2} \times -\frac{1}{2} \times -\frac{1}{2} = -\frac{1}{8}$

Worked exam style question 2

$f(x) = x^3 + px + q$. When $f(x)$ is divided by $(x + 1)$ the remainder is 6.
When $f(x)$ is divided by $(x - 1)$ the remainder is 2. Find the value of p and the value of q.

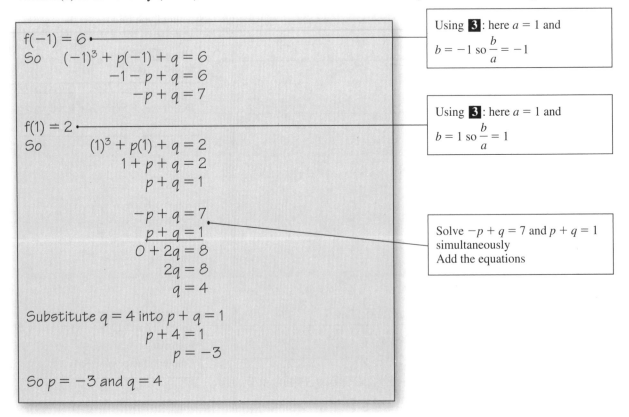

$f(-1) = 6$ ── Using **3**: here $a = 1$ and $b = -1$ so $\frac{b}{a} = -1$

So $(-1)^3 + p(-1) + q = 6$

$-1 - p + q = 6$

$-p + q = 7$

$f(1) = 2$ ── Using **3**: here $a = 1$ and $b = 1$ so $\frac{b}{a} = 1$

So $(1)^3 + p(1) + q = 2$

$1 + p + q = 2$

$p + q = 1$

$-p + q = 7$
$p + q = 1$
$0 + 2q = 8$
$2q = 8$
$q = 4$

Solve $-p + q = 7$ and $p + q = 1$ simultaneously
Add the equations

Substitute $q = 4$ into $p + q = 1$

$p + 4 = 1$

$p = -3$

So $p = -3$ and $q = 4$

Revision exercise 1

1 Simplify $\dfrac{2x^2 - 7x - 4}{x^2 - 16}$.

2 Show that $(x + 1)$ is a factor of $3x^3 - x + 2$.
Hence express $3x^3 - x + 2$ in the form $(x + 1)(Ax^2 + Bx + C)$,
where the values A, B and C are to be found.

3 **(a)** Show that $x - 2$ is a factor of $f(x) = x^3 + 6x^2 - x - 30$.
(b) Hence, or otherwise, find the exact solutions of the equation $f(x) = 0$.

4 Given that -3 is a root of $g(x) = 2x^3 - 5x^2 - 28x + 15$,
find the two positive roots.

5 $f(x) = 2x^3 - 5x^2 - x + a$. The remainder when $f(x)$ is divided by $(x - 1)$ is 2.
(a) Show that $a = 6$.   **(b)** Solve the equation $f(x) = 0$.

6 Show that $9x^3 - 24x^2 - 44x - 16$ is divisible by $(x - 4)$.
Hence express $9x^3 - 24x^2 - 44x - 16$ in the form $(x - 4)(px + q)^2$,
where the values p and q are to be found.

7 $f(x) = x^4 + ax + 5$. When $f(x)$ is divided by $(x + 2)$ the remainder is 7.
(a) Find the value of a.
(b) Find the remainder when $f(x)$ is divided by $(2x - 1)$.

8 **(a)** Factorise $6x^3 - 7x^2 + 1$ completely.
(b) Hence simplify $\dfrac{6x^3 - 7x^2 + 1}{2x^2 - 3x + 1}$.

9 $f(x) = 2x^3 + x^2 + px + q$, where p and q are constants.
Given that $(x - 2)$ and $(x + 3)$ are factors of $f(x)$:
(a) find the values of p and q,
(b) hence solve the equation $f(x) = 0$.

10 Show that $(x - q)$ is a factor of $x^3 + (3 - 2q)x^2 + (q^2 - 6q)x + 3q^2$.

11 $f(x) = 6x^3 + px^2 + qx + 8$, where p and q are constants.
Given that $f(x)$ is exactly divisible by $(x - 2)$, and also that when
$f(x)$ is divided by $(x + 1)$ the remainder is 9:
(a) find the value of p and the value of q,
(b) hence factorise $f(x)$ completely.

12 $f(x) = (x^2 + p)(2x + 3) + 3$, where p is a constant.
(a) Write down the remainder when $f(x)$ is divided by $(2x + 3)$.

Given that the remainder when $f(x)$ is divided by $(x - 2)$ is 24:
(b) prove that $p = -1$,
(c) factorise $f(x)$ completely.

13 $f(n) = n^3 + pn^2 + 11n + 9$, where p is a constant.
(a) Given that $f(n)$ has a remainder of 3 when it is divided
by $(n + 2)$, prove that $p = 6$.
(b) Show that $f(n)$ can be written in the form
$(n + 2)(n + q)(n + r) + 3$, where q and r are
integers to be found.

14 $f(x) = px^3 + 6x^2 + 12x + q$. Given that the remainder when $f(x)$ is divided by $(x - 1)$ is equal to the remainder when $f(x)$ is divided by $(2x + 1)$:

(a) find the value of p.

Given also that $q = 3$, and p has the value found in part (a):

(b) find the value of the remainder.

15 $f(x) = x^3 + ax^2 + bx - 10$, where a and b are constants. When $f(x)$ is divided by $(x - 3)$, the remainder is 14. When $f(x)$ is divided by $(x + 1)$, the remainder is -18.

(a) Find the value of a and the value of b.

(b) Show that $(x - 2)$ is a factor of $f(x)$.

Test yourself	What to review
	If your answer is incorrect
1 Simplify $\dfrac{2x^2 - 5x - 3}{x^2 - x - 6}$.	*Review Heinemann Book C2 pages 1–2* *Revise for C2 page 1* *Example 1*
2 Divide $2x^3 + 7x^2 + 2x - 3$ by $(x + 3)$.	*Review Heinemann Book C2 page 6* *Revise for C2 page 2* *Example 2*
3 Divide $x^3 - 7x + 6$ by $(x - 1)$.	*Review Heinemann Book C2 page 8* *Revise for C2 page 2* *Example 3*
4 Using the factor theorem, show that $(x - 2)$ is a factor of $x^4 - 4x^2 + x - 2$.	*Review Heinemann Book C2 pages 10–11* *Revise for C2 page 2* *Example 3*
5 Factorise $2x^3 + 5x^2 - 4x - 3$ completely.	*Review Heinemann Book C2 page 11* *Revise for C2 page 3* *Example 4*
6 When $3x^3 - x^2 + k$ is divided by $(3x - 1)$ the remainder is 2. Find the value of k.	*Review Heinemann Book C2 page 14* *Revise for C2 page 4* *Worked exam style question 1*

Test yourself answers

1 $\dfrac{2x + 1}{x + 2}$ **2** $2x^2 + x - 1$ **3** $x^2 + x - 6$ **4** $f(2) = 0$ **5** $(x - 1)(x + 3)(2x + 1)$ **6** 2

The sine and cosine rule

Key points to remember

1 The sine rule is

$$\frac{a}{\sin A} = \frac{b}{\sin B} = \frac{c}{\sin C} \quad or \quad \frac{\sin A}{a} = \frac{\sin B}{b} = \frac{\sin C}{c}$$

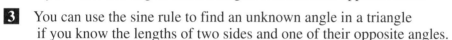

2 You can use the sine rule to find an unknown side in a triangle if you know two angles and the length of one of their opposite sides.

3 You can use the sine rule to find an unknown angle in a triangle if you know the lengths of two sides and one of their opposite angles.

4 There are always two solutions to the equation $\sin x = k$, $0 < x < 180°$, $0 < k < 1$.

$$x = \sin^{-1} k$$
$$and \quad x = 180° - \sin^{-1} k$$

In **3** if the angle you are finding is opposite the larger of the two sides, then there are two possible values for the angle, unless it is a right angle.

5 The cosine rule is:

$$a^2 = b^2 + c^2 - 2bc \cos A$$
$$or \quad b^2 = a^2 + c^2 - 2ac \cos B$$
$$or \quad c^2 = a^2 + b^2 - 2ab \cos C$$

6 You can use the cosine rule to find an unknown side in a triangle if you know the lengths of two sides and the angle between them.

7 You can use the cosine rule to find an unknown angle in a triangle if you know the lengths of all three sides.

8 You can find an unknown angle using a rearranged form of the cosine rule:

$$\cos A = \frac{b^2 + c^2 - a^2}{2bc} \quad or \cos B = \frac{a^2 + c^2 - b^2}{2ac} \quad or \cos C = \frac{a^2 + b^2 - c^2}{2ab}$$

9 You can find the area of a triangle using the formula

$$area = \tfrac{1}{2} ab \sin C \quad or \; area = \tfrac{1}{2} ac \sin B \quad or \; area = \tfrac{1}{2} bc \sin A$$

if you know the lengths of two sides and the value of the angle between them.

Example 1

In each of the following triangles find the value of x.

(a)

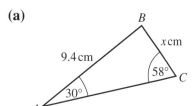

(b)

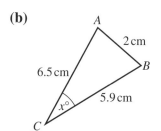

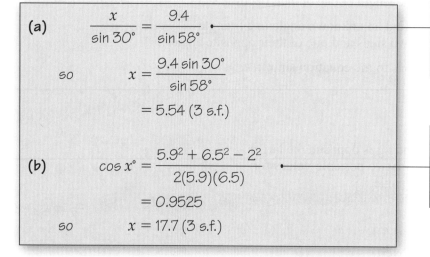

(a)
$$\frac{x}{\sin 30°} = \frac{9.4}{\sin 58°}$$

so
$$x = \frac{9.4 \sin 30°}{\sin 58°}$$

$$= 5.54 \ (3 \text{ s.f.})$$

Using **2**: you have two angles, A and C, and one of the opposite sides, c
Using **1**:
$a = x$, $c = 9.4$, $A = 30°$, $C = 58°$

(b)
$$\cos x° = \frac{5.9^2 + 6.5^2 - 2^2}{2(5.9)(6.5)}$$

$$= 0.9525$$

so
$$x = 17.7 \ (3 \text{ s.f.})$$

Using **7**: you have three sides, so the cosine rule **8** applies:
$\cos C = \dfrac{a^2 + b^2 - c^2}{2ab}$

Example 2

In triangle ABC, $AB = 5$ cm, $BC = 6.4$ cm and angle $ABC = 100°$.
Calculate the area of the triangle, giving your answer to 3 significant figures.

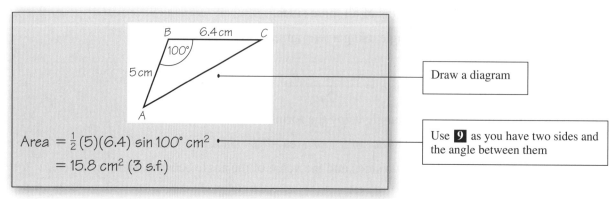

Draw a diagram

$$\text{Area} = \tfrac{1}{2}(5)(6.4) \sin 100° \text{ cm}^2$$

$$= 15.8 \text{ cm}^2 \ (3 \text{ s.f.})$$

Use **9** as you have two sides and the angle between them

Worked exam style question 1

A triangle is to be drawn having two sides of length 10 cm and 12 cm and an area of 42 cm². Calculate the largest possible perimeter that the triangle can have.

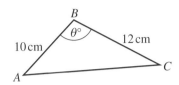

$$42 = \tfrac{1}{2}(10)(12)\sin\theta° \cdot$$

so $\quad \sin\theta° = \dfrac{42}{60} = 0.7$

$$\theta = 44.4 \text{ or } 135.6 \cdot$$

$$AC^2 = 10^2 + 12^2 - 2(10)(12)\cos 135.6° \cdot$$

$$= 415.39 \ldots$$

$$AC = 20.4 \text{ cm (3 s.f.)}$$

Largest perimeter $= (22 + 20.4)$ cm $= 42.4$ cm

Use the area formula **9** to find the angle between the two sides.

Using **4**: there are two possible answers

AC is largest when θ is largest. Use the cosine rule **5** to find AC, with $\theta = 135.6°$.

The perimeter of the other triangle is 30.5 cm

Worked exam style question 2

The lengths of two sides of a triangle T are $(2x + 1)$ cm and $(3 - 2x)$ cm.
Given that the angle between these two sides is $150°$ and that the area of T is A cm²:

(a) show that $A = 1 - (x - \tfrac{1}{2})^2$,
(b) hence, or otherwise, find the maximum value of A and the value of x for which it occurs.

Using **9**: $\sin 150° = \sin 30° = \tfrac{1}{2}$

(a) $A = \tfrac{1}{2}(2x + 1)(3 - 2x)\sin 150°$

$\quad = \tfrac{1}{4}(-4x^2 + 4x + 3)$

$\quad = -(x^2 - x - \tfrac{3}{4})$

$\quad = -[(x - \tfrac{1}{2})^2 - 1]$

$\quad = 1 - (x - \tfrac{1}{2})^2$

(b) The maximum value of $A = 1$, and occurs when $x = \tfrac{1}{2}$ •

See Heinemann Book C1, Section 2.3, for the method for completing the square.

As $(x - \tfrac{1}{2})^2 \geqslant 0$, for $x \in \mathbb{R}$, then from part **(a)** the maximum $= 1$, when $x - \tfrac{1}{2} = 0$. *Otherwise* solve $\dfrac{dA}{dx} = 0$ for x, and substitute that value into the expression for A to find the maximum A.

Worked exam style question 3

In triangle ABC, $AB = 4$ cm, $BC = 2\sqrt{7}$ cm and $AC = (x + 1)$ cm.

(a) Given that $\angle BAC = 60°$, find the value of x.

(b) Given instead that $\cos \angle BAC = \frac{3}{4}$:

   (i) show that x satisfies the equation $x^2 - 4x - 17 = 0$,

   (ii) hence express x in the form $p + \sqrt{q}$, where p and q are integers to be found.

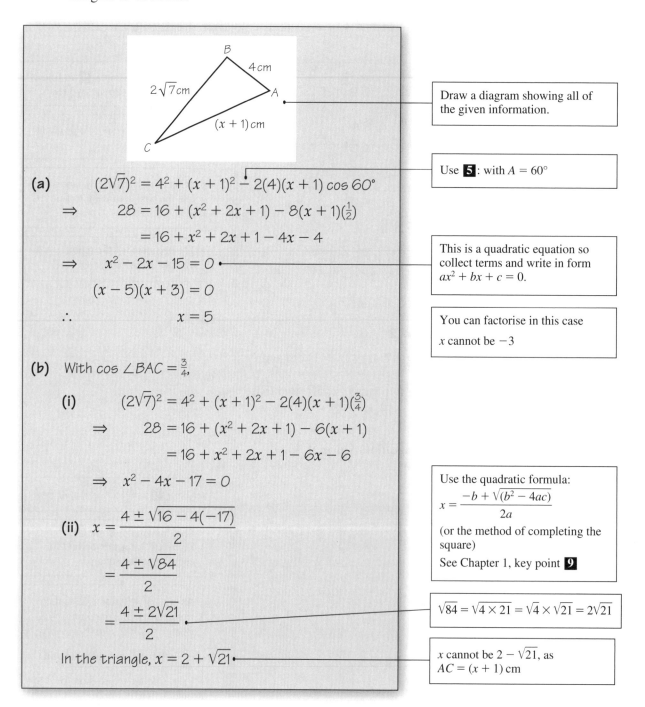

Draw a diagram showing all of the given information.

Use **5**: with $A = 60°$

(a)
$$(2\sqrt{7})^2 = 4^2 + (x + 1)^2 - 2(4)(x + 1)\cos 60°$$
$$\Rightarrow \quad 28 = 16 + (x^2 + 2x + 1) - 8(x + 1)(\tfrac{1}{2})$$
$$= 16 + x^2 + 2x + 1 - 4x - 4$$
$$\Rightarrow \quad x^2 - 2x - 15 = 0$$
$$(x - 5)(x + 3) = 0$$
$$\therefore \qquad x = 5$$

This is a quadratic equation so collect terms and write in form $ax^2 + bx + c = 0$.

You can factorise in this case

x cannot be -3

(b) With $\cos \angle BAC = \frac{3}{4}$,

(i)
$$(2\sqrt{7})^2 = 4^2 + (x + 1)^2 - 2(4)(x + 1)(\tfrac{3}{4})$$
$$\Rightarrow \quad 28 = 16 + (x^2 + 2x + 1) - 6(x + 1)$$
$$= 16 + x^2 + 2x + 1 - 6x - 6$$
$$\Rightarrow \quad x^2 - 4x - 17 = 0$$

(ii)
$$x = \frac{4 \pm \sqrt{16 - 4(-17)}}{2}$$
$$= \frac{4 \pm \sqrt{84}}{2}$$
$$= \frac{4 \pm 2\sqrt{21}}{2}$$

In the triangle, $x = 2 + \sqrt{21}$

Use the quadratic formula:
$$x = \frac{-b + \sqrt{(b^2 - 4ac)}}{2a}$$
(or the method of completing the square)

See Chapter 1, key point **9**

$\sqrt{84} = \sqrt{4 \times 21} = \sqrt{4} \times \sqrt{21} = 2\sqrt{21}$

x cannot be $2 - \sqrt{21}$, as $AC = (x + 1)$ cm

Revision exercise 2

Give answers to 3 significant figures, unless answers are exact.

1 In each of the following diagrams, calculate the value(s) of x.

(a)

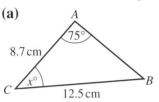

(b)

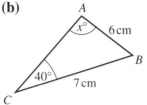

(c)

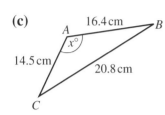

(d)

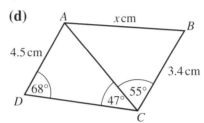

2 In the diagram shown, $AB = 2.8$ cm, $AD = 3.2$ cm, angle $BAD = 30°$, angle $CBD = 48°$ and angle $CDB = 50°$. Calculate the length of BC.

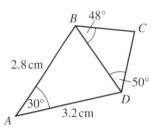

3 From a point A on horizontal ground the angle of elevation of the top of a vertical tower, of height h m, is $25°$; from a point B, 30 m closer to the tower on a straight line between A and the foot of the tower, the angle of elevation is $40°$, as shown in the diagram.
Calculate the value of h.

4 The three sides of a triangular plot of land have lengths of 64 m, 85 m and 120 m. Calculate, in m², the area of this plot.

5 Two different triangles ABC have $AB = 5$ cm, $AC = 3.2$ cm and angle $ABC = 35°$.
Calculate the difference between their areas.

6 In the diagram opposite calculate the value of x and the value of y.

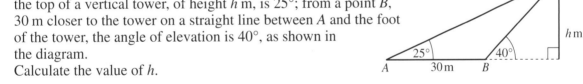

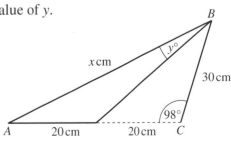

7 In triangle RST, $RS = 4$ cm, $RT = 5$ cm and ST is the longest side. The area and perimeter of the triangle are A cm^2 and P cm respectively. Given that $A = 9$, calculate the value of P.

8 In triangle ABC, $AB = x$ cm, $AC = (x + 1)$ cm, $\angle ABC = 45°$ and $\angle ACB = 30°$. Show that x can be expressed in the form $a + b\sqrt{2}$, where a and b are integers to be found.

9 Use the sine rule in the triangle shown to calculate the value of x.

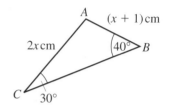

10 A circle has centre O and radius 4 cm. A chord AB subtends an angle of $80°$ at O. The points A and B are joined to a point C on the circumference of the circle, as shown.

Given that $BC = 7.5$ cm, calculate the value of $\angle CAO$.

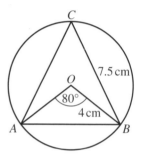

11 The sides of a triangle have lengths x cm, $(x - 2)$ cm and $(x + 1)$ cm. Given that the angle opposite the side of length x is $60°$:
(a) calculate the value of x,
(b) show that the smallest angle in the triangle is approximately $38.2°$.

12 Two sides of a triangle, of area A cm^2, have lengths $(x + 3)$ cm and $(5 - x)$ cm. The angle between these two sides is $30°$.
(a) Show that $A = \frac{1}{4}(15 + 2x - x^2)$.
(b) By using the method of completing the square, or otherwise, find the maximum value of A.

13 The sides of a triangle have lengths 4 cm, $(x - 1)$ cm and $2x$ cm. Given that the angle opposite the side of length $2x$ cm is $120°$:
(a) show that x satisfies the equation $3x^2 - 2x - 13 = 0$, and hence find x,
(b) calculate the value of the smallest angle in the triangle.

14 In triangle ABC, $AC = (x + 2)$ cm, $BC = (x - 2)$ cm and angle $ACB = 60°$.
(a) Show that $AB^2 = x^2 + 12$.
(b) Given that $AB = 6$ cm, show that the area of the triangle is $5\sqrt{3}$ cm^2.

15 The triangle ABC is such that $AB = (x + 4)$ cm, $BC = (3 - x)$ cm and $\angle ABC = 120°$.
(a) Show that $AC^2 = x^2 + x + 37$.
(b) Express AC^2 in the form $(x + a)^2 + b$, where a and b are constants to be found.
(c) Hence find the minimum value of AC, and the value of x for which it occurs.

Test yourself	**What to review**

Give answers to 3 significant figures, unless otherwise instructed.

If your answer is incorrect

1 Find the value of x in the triangle below.

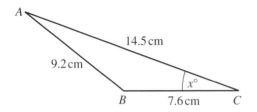

Review Heinemann Book C2 pages 27–28
Revise for C2 page 8
Example 1(b)

---

2 In $\triangle ABC$, $AB = 3$ cm, $AC = 4$ cm and $\angle BAC = 115°$.
Calculate:

(a) the area of the triangle,

(b) the length of BC.

(a) Review Heinemann Book C2 pages 32–33
Revise for C2 page 8
Example 2

(b) Review Heinemann Book C2 pages 24–25
Revise for C2 page 9
Worked exam style question 1

---

3 The area of triangle ABC is 3 cm². The longest side $AB = 5\sqrt{2}$ cm, and $BC = \sqrt{2}$ cm.

(a) Show that $AC = 6$ cm.

(b) Given that the smallest angle is $x°$, find the value of $\sin x°$ in the form $k\sqrt{2}$.

(a) Review Heinemann Book C2 pages 32–33
Revise for C2 page 9
Worked exam style question 1

(b) Review Heinemann Book C2 page 21
Revise for C2 page 8
Example 1(a)

---

4 Two sides of a triangle, of area A cm², have lengths $(2 - x)$ cm and $(2x + 1)$ cm and the angle between the two sides is 30°, as shown.

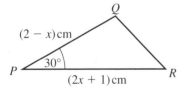

Review Heinemann Book C2 pages 32–33
Revise for C2 page 9
Worked exam style question 2

(a) Show that A can be given as $A = \frac{1}{4}(2 + 3x - 2x^2)$.

(b) By using the method of completing the square, or otherwise, find the maximum value of A and the value of x for which it occurs.

Test yourself answers

1 33.7°   **2 (a)** 5.44 cm²   **(b)** 5.93 cm   **3 (a)** $\sin ABC = 0.6$ using , then use cosine rule   **(b)** $\dfrac{\sqrt{2}}{10}$ **4 (b)** maximum $A = \frac{25}{32}$, when $x = \frac{3}{4}$

Exponentials and logarithms

3

Key points to remember

1 A function $y = a^x$, or $f(x) = a^x$, where a is a constant, is called an exponential function.

2 $\log_a n = x$ means that $a^x = n$, where a is called the base of the logarithm.

3 $\log_a 1 = 0$
$\log_a a = 1$

4 $\log_{10} x$ is sometimes written as $\log x$.

5 The laws of logarithms are:

$\log_a xy = \log_a x + \log_a y$ \qquad (the multiplication law)

$\log_a \left(\dfrac{x}{y}\right) = \log_a x - \log_a y$ \qquad (the division law)

$\log_a (x)^k = k \log_a x$ \qquad (the power law)

6 From the power law,

$\log_a \left(\dfrac{1}{x}\right) = -\log_a x.$

7 You can solve an equation such as $a^x = b$ by first taking logarithms (to base 10) of each side.

8 The change of base rule for logarithms can be written as

$\log_a x = \dfrac{\log_b x}{\log_b a}.$

9 From the change of base rule, $\log_a b = \dfrac{1}{\log_b a}.$

Example 1

(a) On the same axes, sketch the graph of $y = 4^x$ and the graph of $y = (\frac{1}{4})^x$.

(b) Write down the value of $\log_4 16$ and the value of $\log_{\frac{1}{4}} 16$.

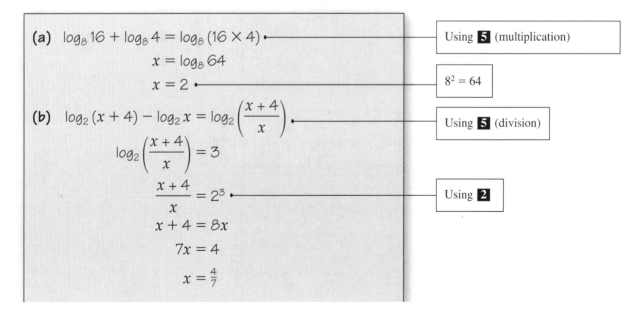

(a)

$y = \left(\frac{1}{4}\right)^x$ $y = 4^x$

> The graph of $y = a^x$ always passes through the point $(0, 1)$

> $\left(\frac{1}{4}\right)^x = (4^{-1})^x = 4^{-x}$
> The graph of $y = \left(\frac{1}{4}\right)^x$ is a reflection in the y-axis of the graph of $y = 4^x$

> Using **2**

(b) $\log_4 16 = 2$

$\log_{\left(\frac{1}{4}\right)} 16 = -2$

> $\left(\frac{1}{4}\right)^{-2} = \dfrac{1}{\left(\frac{1}{4}\right)^2}$
>
> $= \dfrac{1}{\frac{1}{16}} = 16$

Example 2

Find the value of x for which:

(a) $\log_8 x = 1$ **(b)** $\log_x 10\,000 = 2$ **(c)** $\log_x 27 = x$.

(a) $x = 8$

(b) $x = 100$

(c) $x = 3$

> Using **2** and **3**:
> $\log_a a = 1$ $(8^1 = 8)$
> $100^2 = 10\,000$
> $3^3 = 27$

Example 3

Solve each equation to find the value of x:

(a) $x = \log_8 16 + \log_8 4$ **(b)** $\log_2 (x + 4) - \log_2 x = 3$ **(c)** $2\log_5 x - \log_5 3 = \log_5 (x + 6)$.

(a) $\log_8 16 + \log_8 4 = \log_8 (16 \times 4)$

$x = \log_8 64$

$x = 2$

> Using **5** (multiplication)

> $8^2 = 64$

(b) $\log_2 (x + 4) - \log_2 x = \log_2 \left(\dfrac{x + 4}{x}\right)$

$\log_2 \left(\dfrac{x + 4}{x}\right) = 3$

$\dfrac{x + 4}{x} = 2^3$

$x + 4 = 8x$

$7x = 4$

$x = \frac{4}{7}$

> Using **5** (division)

> Using **2**

(c) $2\log_5 x - \log_5 3 = \log_5(x+6)$

$\qquad\qquad 2\log_5 x = \log_5(x^2)$ — Using **5** (power)

$\qquad \log_5(x^2) - \log_5 3 = \log_5\left(\dfrac{x^2}{3}\right)$ — Using **5** (division)

$\qquad\qquad \log_5\left(\dfrac{x^2}{3}\right) = \log_5(x+6)$

$\qquad\qquad\qquad \dfrac{x^2}{3} = x+6$

$\qquad\qquad\qquad x^2 = 3x + 18$

$\qquad\quad x^2 - 3x - 18 = 0$

$\qquad\quad (x-6)(x+3) = 0$

$\qquad\qquad x = 6 \ (x = -3 \text{ is not valid})$ — $\log_5(-3)$ does not exist

Worked exam style question 1

Given that $\log_a x = p$ and $\log_a y = q$, find in terms of p and q:

(a) $\log_a(ax^2 y)$ $\qquad$ **(b)** $\log_a\left(\dfrac{y^3}{a^2}\right)$.

(a) $\log_a(ax^2 y) = \log_a a + \log_a(x^2) + \log_a y$ — Using **5** (multiplication)

$\qquad\qquad\qquad = \log_a a + 2\log_a x + \log_a y$ — Using **5** (power)

$\qquad\qquad\qquad = 1 + 2p + q$ — Using **3**

(b) $\log_a\left(\dfrac{y^3}{a^2}\right) = \log_a(y^3) - \log_a(a^2)$ — Using **5** (division)

$\qquad\qquad\qquad = 3\log_a y - 2\log_a a$ — Using **5** (power)

$\qquad\qquad\qquad = 3q - 2$ — Using **3**

Worked exam style question 2

Solve the equation $3^{1+2x} - 3^x = 4$, giving your answer to 3 significant figures.

Let $y = 3^x$

$\qquad 3^{1+2x} = 3^1 \times 3^{2x} = 3 \times (3^x)^2 = 3y^2$

So the equation becomes: $3y^2 - y - 4 = 0$

$\qquad\qquad\qquad (3y-4)(y+1) = 0$

$\qquad\qquad\qquad y = \tfrac{4}{3} \text{ or } y = -1$

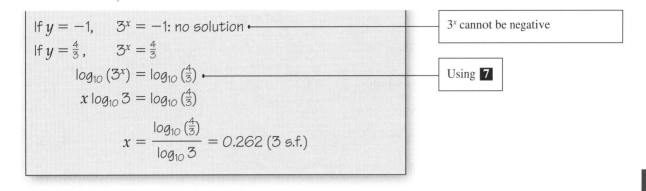

If $y = -1$, $\quad 3^x = -1$: no solution ← 3ˣ cannot be negative

3^x cannot be negative

If $y = \frac{4}{3}$, $\quad 3^x = \frac{4}{3}$

$\log_{10}(3^x) = \log_{10}\left(\frac{4}{3}\right)$ → Using **7**

$x \log_{10} 3 = \log_{10}\left(\frac{4}{3}\right)$

$x = \dfrac{\log_{10}\left(\frac{4}{3}\right)}{\log_{10} 3} = 0.262$ (3 s.f.)

Worked exam style question 3

Given that $\log_8 x + \log_2 x = 12 \log_x 2$, find the value of x.

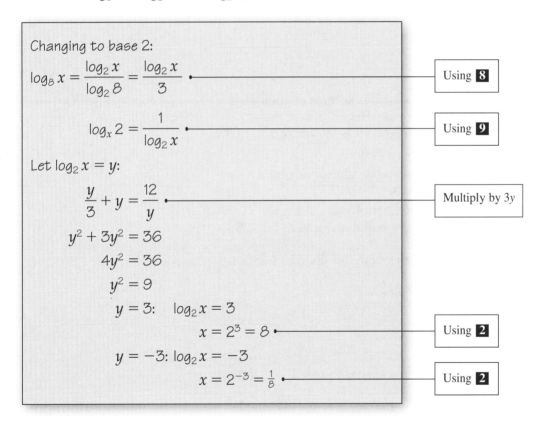

Changing to base 2:

$\log_8 x = \dfrac{\log_2 x}{\log_2 8} = \dfrac{\log_2 x}{3}$ → Using **8**

$\log_x 2 = \dfrac{1}{\log_2 x}$ → Using **9**

Let $\log_2 x = y$:

$\dfrac{y}{3} + y = \dfrac{12}{y}$ → Multiply by $3y$

$y^2 + 3y^2 = 36$

$4y^2 = 36$

$y^2 = 9$

$y = 3$: $\quad \log_2 x = 3$

$x = 2^3 = 8$ → Using **2**

$y = -3$: $\log_2 x = -3$

$x = 2^{-3} = \frac{1}{8}$ → Using **2**

Revision exercise 3

1 Find the value of:

   (a) $\log_{10} 10\,000$ **(b)** $\log_4 8$ **(c)** $\log_5 0.2$.

2 Solve each equation to find the value of x:

   (a) $\log_4 (x - 6) + \log_4 x = 2$ **(b)** $\log_x 9 - \log_x 4 = 2$.

3 Given that $\log_2 x = m$ and $\log_2 y = n$, express in terms of m and n:

(a) $\log_2 (8x^3y)$ (b) $\log_2 \left(\dfrac{y}{2x}\right)$.

Given also that $\log_2 (8x^3y) = 20$ and $\log_2 \left(\dfrac{y}{2x}\right) = 0$:

(c) find the value of m and the value of n,

(b) find the value of x and the value of y.

4 Find, to 3 significant figures, the value of x for which $7^x = 77$.

5 Solve for x, giving your answers to 2 decimal places:

$6^{2x} - 16(6^x) + 48 = 0.$

6 Find, to 3 significant figures, the value of $\log_{11} 200$.

7 Solve the simultaneous equations:

$\log_2 x + \log_8 y = -1$
$\log_4 x + \log_2 y = 2.$

8 Every £1 of money invested in a savings scheme continuously gains interest at a rate of 4% per year. Hence, after x years, the total value of an initial £1 investment is £y, where $y = 1.04^x$.

(a) Sketch the graph of $y = 1.04^x$, $x \geqslant 0$.

(b) Calculate, to the nearest £, the total value of an initial £800 investment after 10 years.

(c) Use logarithms to find the number of years it takes to double the total value of any initial investment.

9 Given that $\log_2 x = a$, find, in terms of a, the simplest form of:

(a) $\log_2 (16x)$ (b) $\log_2 \left(\dfrac{x^4}{2}\right)$.

(c) Hence, or otherwise, solve

$$\log_2 16x - \log_2 \left(\frac{x^4}{2}\right) = \frac{1}{2}$$

giving your answer in its simplest surd form.

10 Given that $\log_5 x = a$ and $\log_5 y = b$, find, in terms of a and b:

(a) $\log_5 \left(\dfrac{x^2}{y}\right)$ (b) $\log_5 (25x\sqrt{y})$.

It is given that $\log_5 \left(\dfrac{x^2}{y}\right) = 1$ and that $\log_5 (25x\sqrt{y}) = 1$.

(c) Form simultaneous equations in a and b.

(d) Show that $a = -0.25$ and find the value of b.

Using the values of a and b, or otherwise:

(e) calculate, to 3 decimal places, the value of x and the value of y.

Test yourself	What to review
	If your answer is incorrect
1 Given that $\log_3 x = p$ and $\log_3 y = q$, express in terms of p and q: (a) $\log_3 (x^2\sqrt{y})$ (b) $\log_3 \left(\dfrac{9y^4}{x}\right)$.	*Review Heinemann Book C2 pages 41–42* *Revise for C2 page 16* *Worked exam style question 1*
2 Solve each equation to find the value of x: (a) $4 \log_x 3 + 2 \log_x 2 = 2$ (b) $\log_5 (3x + 1) - \log_5 (2x) = 1$.	*Review Heinemann Book C2 pages 41–42* *Revise for C2 page 15* *Example 3*
3 Find, to 3 significant figures, the value of x for which: $3^{2x+1} = 105$.	*Review Heinemann Book C2 pages 43–44* *Revise for C2 page 16* *Worked exam style question 2*
4 Solve, giving your answer to 3 significant figures: $2^{2x+1} - 2^x - 21 = 0$.	*Review Heinemann Book C2 pages 43–44* *Revise for C2 page 16* *Worked exam style question 2*
5 Find the values of x for which: $\log_3 x + 2 \log_x 3 = 3$.	*Review Heinemann Book C2 pages 45–46* *Revise for C2 page 17* *Worked exam style question 3*
6 Given that $\log_2 (x + 1) = \log_4 (7x - 5)$: (a) find the possible values of x, (b) evaluate $\log_2 (x + 1)$ for each of the x-values found in part (a).	*Review Heinemann Book C2 pages 45–46* *Revise for C2 page 17* *Worked exam style question 3*

Test yourself answers

5 $x = 3, x = 9$ **6** (a) $x = 2, x = 3$ (b) 1.58 (3 s.f.) and 2

1 (a) $2p + \frac{1}{2}q$ (b) $2 + 4q - p$ **2** (a) 18 (b) $\frac{4}{3}$ **3** 1.62 **4** 1.81

Coordinate geometry in the (x, y) plane

4

Key points to remember

1 The mid-point of (x_1, y_1) and (x_2, y_2) is
$$\left(\frac{x_1 + x_2}{2}, \frac{y_1 + y_2}{2}\right).$$

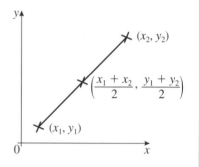

2 The distance d between (x_1, y_1) and (x_2, y_2) is
$d = \sqrt{[(x_2 - x_1)^2 + (y_2 - y_1)^2]}.$

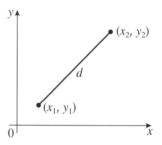

3 The equation of the circle centre (a, b) radius r is
$(x - a)^2 + (y - b)^2 = r^2.$

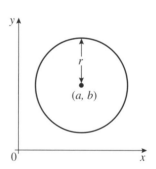

4 A chord is a line that joins two points on the circumference of a circle.

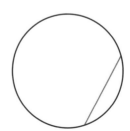

5 The perpendicular from the centre of a circle to a chord bisects the chord.

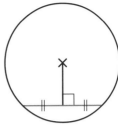

6 The angle in a semicircle is a right angle.

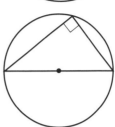

7 A tangent is a line that meets a circle at one point only.

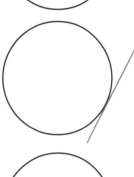

8 The angle between a tangent and a radius is 90°.

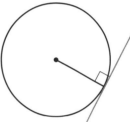

Example 1

Work out the coordinates of the centre and the radius of the circle
$x^2 + y^2 - 12x + 10y - 60 = 0$.

$$x^2 + y^2 - 12x + 10y - 60 = 0$$
$$x^2 - 12x + y^2 + 10y - 60 = 0$$
$$(x - 6)^2 - 6^2 + (y + 5)^2 - 5^2 - 60 = 0$$
$$(x - 6)^2 + (y + 5)^2 - 60 - 6^2 - 5^2 = 0$$
$$(x - 6)^2 + (y + 5)^2 - 60 - 36 - 25 = 0$$
$$(x - 6)^2 + (y + 5)^2 - 121 = 0$$
$$(x - 6)^2 + (y + 5)^2 = 121$$
$$(x - 6)^2 + (y + 5)^2 = 11^2$$

So the centre is $(6, -5)$ and the radius $= 11$

Rearrange the equation into the form
$(x - a)^2 + (y - b)^2 = r^2$

Collect the terms in x and the terms in y together

Complete the square:

use $x^2 - 2ax = (x - a)^2 - a^2$
here $a = 6$, so $x^2 - 12x = (x - 6)^2 - 6^2$
$y^2 + 2by = (y + b)^2 - b^2$
here $b = 5$, so $y^2 + 10y = (y + 5)^2 - 5^2$

Simplify

Using **3**: here $(a, b) = (6, -5)$ and $r = 11$

Example 2

Work out the coordinates of the points where the circle $(x - 5)^2 + (y - 2)^2 = 34$ meets the y-axis.

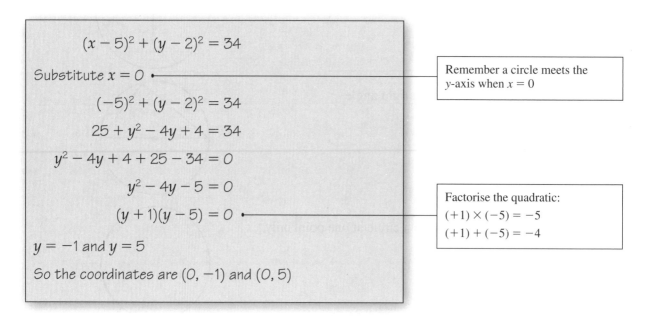

$$(x - 5)^2 + (y - 2)^2 = 34$$

Substitute $x = 0$

$$(-5)^2 + (y - 2)^2 = 34$$
$$25 + y^2 - 4y + 4 = 34$$
$$y^2 - 4y + 4 + 25 - 34 = 0$$
$$y^2 - 4y - 5 = 0$$
$$(y + 1)(y - 5) = 0$$

$y = -1$ and $y = 5$

So the coordinates are $(0, -1)$ and $(0, 5)$

Remember a circle meets the y-axis when $x = 0$

Factorise the quadratic:
$(+1) \times (-5) = -5$
$(+1) + (-5) = -4$

Example 3

The point $P(-1, 6)$ lies on the circle centre $(3, 4)$.

(a) Work out the radius of the circle.

(b) Hence, write down an equation of the circle.

(c) Find an equation of the tangent to the circle at P.

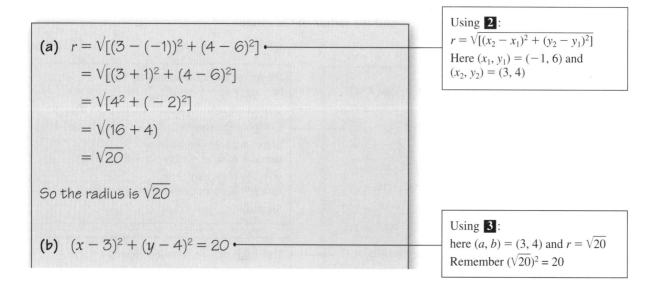

(a) $r = \sqrt{[(3 - (-1))^2 + (4 - 6)^2]}$
$\quad = \sqrt{[(3 + 1)^2 + (4 - 6)^2]}$
$\quad = \sqrt{[4^2 + (-2)^2]}$
$\quad = \sqrt{(16 + 4)}$
$\quad = \sqrt{20}$

So the radius is $\sqrt{20}$

(b) $(x - 3)^2 + (y - 4)^2 = 20$

Using **2**:
$r = \sqrt{[(x_2 - x_1)^2 + (y_2 - y_1)^2]}$
Here $(x_1, y_1) = (-1, 6)$ and $(x_2, y_2) = (3, 4)$

Using **3**:
here $(a, b) = (3, 4)$ and $r = \sqrt{20}$
Remember $(\sqrt{20})^2 = 20$

(c) Gradient of radius $= \dfrac{4-6}{3-(-1)}$

$$= \dfrac{4-6}{3+1}$$

$$= \dfrac{-2}{4}$$

$$= -\dfrac{1}{2}$$

Using $m = \dfrac{y_2 - y_1}{x_2 - x_1}$

Here $(x_1, y_1) = (-1, 6)$ and $(x_2, y_2) = (3, 4)$

So gradient of tangent is $\dfrac{-1}{\left(-\frac{1}{2}\right)} = 2$

Using **8** and $\dfrac{-1}{m}$

Remember $\dfrac{1}{\left(\frac{a}{b}\right)} = \dfrac{b}{a}$

$$y - 6 = 2(x - (-1))$$
$$y - 6 = 2(x + 1)$$
$$y - 6 = 2x + 2$$
$$y = 2x + 8$$

So the equation of the tangent at P is $y = 2x + 8$

Using $y - y_1 = m(x - x_1)$
Here $(x_1, y_1) = (-1, 6)$ and $m = 2$

Worked exam style question 1

Work out the coordinates of the points where the line $y = x + 10$ meets the circle $(x - 2)^2 + (y - 6)^2 = 68$.

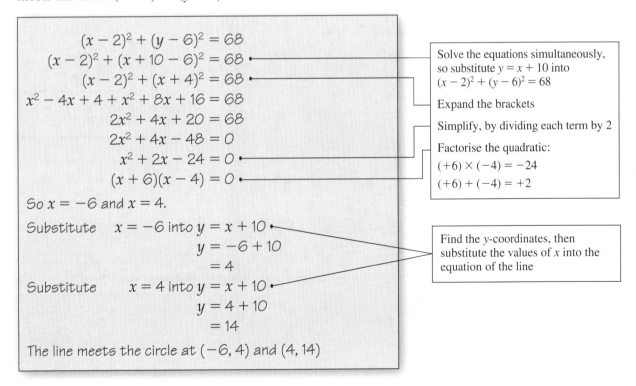

$$(x - 2)^2 + (y - 6)^2 = 68$$
$$(x - 2)^2 + (x + 10 - 6)^2 = 68$$
$$(x - 2)^2 + (x + 4)^2 = 68$$
$$x^2 - 4x + 4 + x^2 + 8x + 16 = 68$$
$$2x^2 + 4x + 20 = 68$$
$$2x^2 + 4x - 48 = 0$$
$$x^2 + 2x - 24 = 0$$
$$(x + 6)(x - 4) = 0$$

So $x = -6$ and $x = 4$.

Substitute $x = -6$ into $y = x + 10$
$$y = -6 + 10$$
$$= 4$$

Substitute $x = 4$ into $y = x + 10$
$$y = 4 + 10$$
$$= 14$$

The line meets the circle at $(-6, 4)$ and $(4, 14)$

Solve the equations simultaneously, so substitute $y = x + 10$ into $(x - 2)^2 + (y - 6)^2 = 68$

Expand the brackets

Simplify, by dividing each term by 2

Factorise the quadratic:
$(+6) \times (-4) = -24$
$(+6) + (-4) = +2$

Find the y-coordinates, then substitute the values of x into the equation of the line

Worked exam style question 2

A circle has centre $C(-2, 0)$ and radius 5. A tangent is drawn from the point $P(7, 4)$ to touch the circle at T. Find the length of PT.

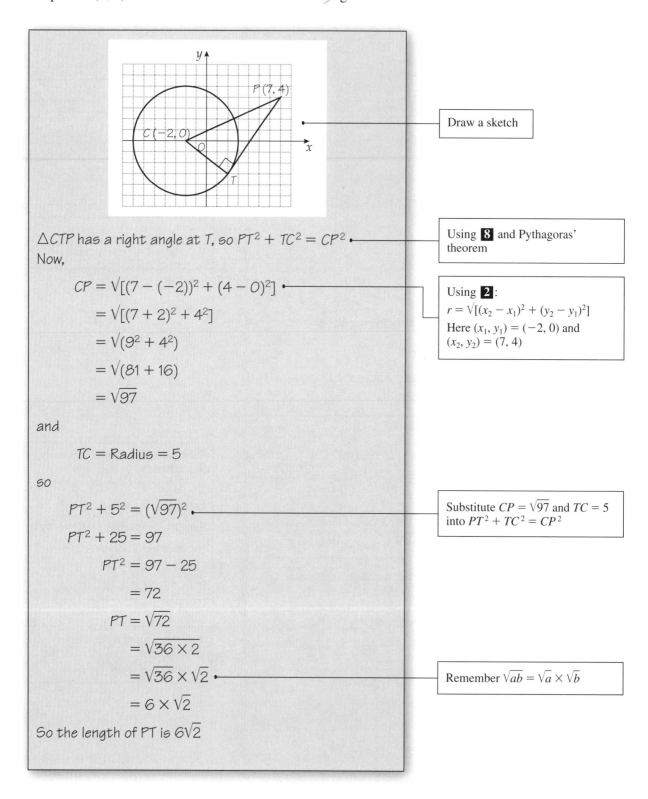

Draw a sketch

$\triangle CTP$ has a right angle at T, so $PT^2 + TC^2 = CP^2$

Now,

Using **8** and Pythagoras' theorem

$$CP = \sqrt{[(7 - (-2))^2 + (4 - 0)^2]}$$

$$= \sqrt{[(7 + 2)^2 + 4^2]}$$

$$= \sqrt{(9^2 + 4^2)}$$

$$= \sqrt{(81 + 16)}$$

$$= \sqrt{97}$$

Using **2**:
$r = \sqrt{[(x_2 - x_1)^2 + (y_2 - y_1)^2]}$
Here $(x_1, y_1) = (-2, 0)$ and $(x_2, y_2) = (7, 4)$

and

$$TC = \text{Radius} = 5$$

so

$$PT^2 + 5^2 = (\sqrt{97})^2$$

$$PT^2 + 25 = 97$$

$$PT^2 = 97 - 25$$

$$= 72$$

$$PT = \sqrt{72}$$

$$= \sqrt{36 \times 2}$$

$$= \sqrt{36} \times \sqrt{2}$$

$$= 6 \times \sqrt{2}$$

So the length of PT is $6\sqrt{2}$

Substitute $CP = \sqrt{97}$ and $TC = 5$ into $PT^2 + TC^2 = CP^2$

Remember $\sqrt{ab} = \sqrt{a} \times \sqrt{b}$

Revision exercise 4

1 (a) Work out the coordinates of the centre and the radius of the circle $x^2 + y^2 - 2x - 8y + 1 = 0$.

(b) Draw a sketch of the circle.

2 (a) Sketch the circle $(x + 3)^2 + (y - 7)^2 = 58$.

(b) Work out the coordinates of the points where the circle meets the coordinate axes.

3 The circle $(x - 4)^2 + (y - 5)^2 = 41$ meets the coordinate axes at $A(a, 0)$ and $B(0, b)$, where $a > 0$ and $b > 0$.

(a) Work out the value of a and the value of b.

(b) Work out the area of $\triangle AOB$, where O is the origin.

4 The point $A(-3, 4)$ lies on the circle centre $(2, -1)$.

(a) Work out the radius of the circle.

(b) Hence write down an equation of the circle.

(c) Find an equation of the tangent to the circle at A.

5 The points $P(6, 0)$, $Q(-2, 2)$ and $R(a, 8)$ lie on the circle centre $(3, 5)$. The line QR is a diameter of the circle.

(a) Work out the value of a.

(b) Show that $\triangle PQR$ has a right angle.

6 The points $A(-4, -2)$, $B(-8, 6)$, $C(0, 10)$ and $D(4, 2)$ lie on a circle. $ABCD$ is a square.

(a) Work out the area of the square.

(b) Find an equation for the circle.

7 The point $P(1, 9)$ lies on the circle $(x + 2)^2 + (y - 3)^2 = r^2$.

(a) Find the value of r. Write your answer in the form $p\sqrt{q}$, where the values of p and q are to be found.

(b) Find an equation for the tangent to the circle at P.

8 The line $y = x - 5$ meets the circle $(x - 1)^2 + (y + 2)^2 = 34$ at the points A and B.

(a) Work out the coordinates of A and B.

(b) Work out the length of the chord AB.

9 (a) Sketch the circle $(x + 4)^2 + (y - 5)^2 = 9$.

(b) Work out the exact length of the tangents from the point $(3, 5)$ to the circle.

10 The line $4x + 3y = 24$ meets the coordinate axes at $A(a, 0)$ and $B(0, b)$. The line AB is a diameter of a circle.

(a) Work out the values of a and b.

(b) Find the equation of the circle.

(c) Show that the circle passes through the origin.

11 The points $P(-1, -8)$, $Q(5, 6)$ and $R(9, 2)$ lie on a circle.

  (a) Show that $\triangle PQR$ has a right angle.

  (b) Work out the coordinates of the centre of the circle.

  (c) Find an equation of the circle.

12 The line $x - 2y - 9 = 0$ meets the circle $x^2 + (y - 3)^2 = 45$ at P.
The line l is parallel to $x - 2y - 9 = 0$ and meets the circle at Q.

  (a) Work out the coordinates of P.

  (b) Find an equation of l.

13 The line $y = 2x + 5$ meets the circle $(x + 1)^2 + (y - 3)^2 = 80$
at the points A and B.

  (a) Work out the coordinates of A and B.

  (b) Show that the chord AB is a diameter of the circle.

14 The line $y = x + 1$ meets the circle $x^2 + y^2 - 4x - 8y - 5 = 0$
at the points S and T.

  (a) Show that the coordinates of the centre of the circle are
(2, 4) and find the radius.

  (b) Work out the coordinates of S and T.

  (c) Find the shortest distance between the centre of the circle
and the line.

15 The line MN is a chord of the circle $(x - 4)^2 + (y - 2)^2 = n$.
The coordinates of M and N are $(0, -4)$ and $(-2, 6)$
respectively.

  (a) Work out the value of n.

  The line l is perpendicular to MN and bisects it.

  (b) Find an equation for l.

  (c) Show that l passes through the centre of the circle.

16 The circle C has centre $(5, 13)$ and touches the x-axis.

  (a) Find an equation of C in terms of x and y.

  (b) Find an equation of the tangent to C at the point $(10, 1)$,
giving your answer in the form $ay + bx + c = 0$,
where a, b and c are integers. **E**

17 The circle C, with centre (a, b) and radius 5, touches the
x-axis at $(4, 0)$, as shown in the diagram.

  (a) Write down the value of a and the value of b.

  (b) Find an equation of C.

  A tangent to the circle, drawn from the point $P(8, 17)$,
touches the circle at T.

  (c) Find, to 3 significant figures, the length of PT. **E**

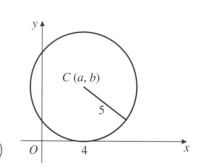

18 The point A has coordinates $(2, 5)$ and the point B has coordinates $(-2, 8)$. Find an equation of the circle with diameter AB.

19 A circle C, has equation $x^2 + y^2 - 6x + 8y - 75 = 0$.

   (a) Write down the coordinates of the centre of C, and calculate the radius of C.

   A second circle has centre at the point $(15, 12)$ and radius 10.

   (b) Sketch both circles on a single diagram and find the coordinates of the point where they touch.

20 The circle C has equation $x^2 + y^2 - 8x - 16y - 209 = 0$.

   (a) Find the coordinates of the centre of C and the radius of C.

   (b) Find an equation of the tangent to C at the point $(21, 8)$.

Test yourself	What to review
	If your answer is incorrect
1 The line PQ is a diameter of a circle, where P and Q are $(-4, 2)$ and $(6, 8)$ respectively. Work out the coordinates of the centre of the circle.	*Review Heinemann Book C2 page 50*
2 Work out the exact distance between the points $(4, -3)$ and $(3, 1)$.	*Review Heinemann Book C2 page 57* *Revise for C2 page 22 Example 3*
3 Write down an equation of the circle centre $(7, -5)$ and radius 6.	*Review Heinemann Book C2 page 60* *Revise for C2 page 22 Example 3*
4 The line $y = \frac{1}{2}x + \frac{3}{2}$ touches the circle $(x + 1)^2 + (y - 6)^2 = 20$ at $P(1, 2)$. Show that the radius at P is perpendicular to the line.	*Review Heinemann Book C2 pages 62–63*
5 Work out where the circle $(x - 5)^2 + (y + 1)^2 = 9$ meets the x-axis.	*Review Heinemann Book C2 page 64* *Revise for C2 page 22 Example 2*
6 Show that the line $y = 3 - x$ does not meet the circle $x^2 + (y + 2)^2 = 8$.	*Review Heinemann Book C2 page 65*

Test yourself answers

1 $(1, 5)$ **2** $\sqrt{17}$ **3** $(x - 7)^2 + (y + 5)^2 = 36$ **4** $-2 \times \frac{1}{2} = -1$ **5** $(5 + 2\sqrt{2}, 0), (5 - 2\sqrt{2}, 0)$ **6** $b^2 - 4ac < 0$

The binomial expansion

5

Key points to remember

1 You can use Pascal's Triangle to multiply out a bracket.

2 You can use combinations and factional notation to help you expand binomial expressions. For larger indices it is quicker than using Pascal's Triangle.

3 $n! = n \times (n-1) \times (n-2) \times (n-3) \times \ldots \times 3 \times 2 \times 1$

4 The number of ways of choosing r items from a group of n items is written $^{n}C_r$ or $\binom{n}{r}$.

e.g. $^{3}C_2 = \dfrac{3!}{(3-2)!2!} = \dfrac{6}{1 \times 2} = 3$

5 The binomial expansion is

$$(a+b)^n = {}^{n}C_0 a^n + {}^{n}C_1 a^{n-1}b + {}^{n}C_2 a^{n-2}b^2 + {}^{n}C_3 a^{n-3}b^3 + \ldots + {}^{n}C_n b^n$$

or $\binom{n}{0}a^n + \binom{n}{1}a^{n-1}b + \binom{n}{2}a^{n-2}b^2 + \binom{n}{3}a^{n-3}b^3 + \ldots + \binom{n}{n}b^n$

6 Similarly,

$$(a+bx)^n = {}^{n}C_0 a^n + {}^{n}C_1 a^{n-1}bx + {}^{n}C_2 a^{n-2}b^2x^2 + {}^{n}C_3 a^{n-3}b^3x^3 + \ldots {}^{n}C_n b^n x^n$$

or $\binom{n}{0}a^n + \binom{n}{1}a^{n-1}bx + \binom{n}{2}a^{n-2}b^2x^2 + \binom{n}{3}a^{n-3}b^3x^3 + \ldots + \binom{n}{n}b^n x^n$

7 $(1+x)^n = 1 + nx + \dfrac{n(n-1)}{2!}x^2 + \dfrac{n(n-1)(n-2)}{3!}x^3 + \dfrac{n(n-1)(n-2)(n-3)}{4!}x^4 + \ldots$

8 To expand $(a+x)^n$ using **7**, take out a factor of a: $(a+x)^n = \left(a\left(1 + \dfrac{x}{a}\right) \right)^n = a^n\left(1 + \dfrac{x}{a}\right)^n$.

Example 1

Use Pascal's Triangle to find the expansion of $(2x - 3y)^4$.

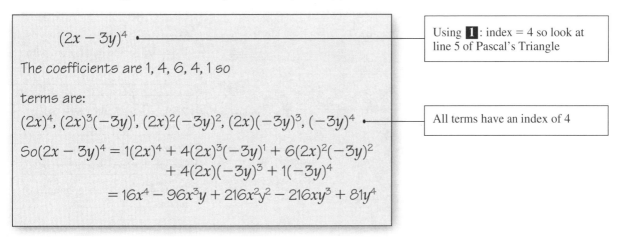

$(2x - 3y)^4$ •──────────────── | Using **1**: index = 4 so look at line 5 of Pascal's Triangle

The coefficients are 1, 4, 6, 4, 1 so

terms are:

$(2x)^4, (2x)^3(-3y)^1, (2x)^2(-3y)^2, (2x)(-3y)^3, (-3y)^4$ •─── | All terms have an index of 4

So $(2x - 3y)^4 = 1(2x)^4 + 4(2x)^3(-3y)^1 + 6(2x)^2(-3y)^2$
$+ 4(2x)(-3y)^3 + 1(-3y)^4$
$= 16x^4 - 96x^3y + 216x^2y^2 - 216xy^3 + 81y^4$

Example 2

The coefficient of x^2 in the expansion of $(1 + x)(3 + ax)^3$ is 10.
Find the possible values of a.

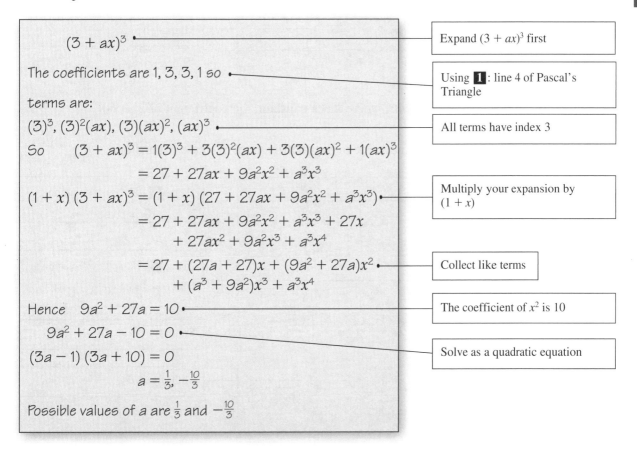

$(3 + ax)^3$ •──────────────── | Expand $(3 + ax)^3$ first

The coefficients are 1, 3, 3, 1 so •─── | Using **1**: line 4 of Pascal's Triangle

terms are:

$(3)^3, (3)^2(ax), (3)(ax)^2, (ax)^3$ •─── | All terms have index 3

So $(3 + ax)^3 = 1(3)^3 + 3(3)^2(ax) + 3(3)(ax)^2 + 1(ax)^3$
$= 27 + 27ax + 9a^2x^2 + a^3x^3$

$(1 + x)(3 + ax)^3 = (1 + x)(27 + 27ax + 9a^2x^2 + a^3x^3)$ •─── | Multiply your expansion by $(1 + x)$
$= 27 + 27ax + 9a^2x^2 + a^3x^3 + 27x$
$+ 27ax^2 + 9a^2x^3 + a^3x^4$
$= 27 + (27a + 27)x + (9a^2 + 27a)x^2$ •─── | Collect like terms
$+ (a^3 + 9a^2)x^3 + a^3x^4$

Hence $9a^2 + 27a = 10$ •─── | The coefficient of x^2 is 10

$9a^2 + 27a - 10 = 0$ •───

$(3a - 1)(3a + 10) = 0$ | Solve as a quadratic equation

$a = \frac{1}{3}, -\frac{10}{3}$

Possible values of a are $\frac{1}{3}$ and $-\frac{10}{3}$

Example 3

Use the binomial expansion to find the first three terms in the expansions of:

(a) $(1 - 3x)^7$ **(b)** $(2 + 5x)^5$.

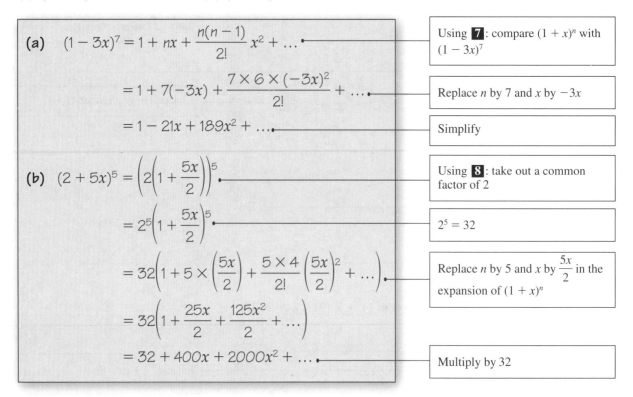

(a) $(1 - 3x)^7 = 1 + nx + \dfrac{n(n - 1)}{2!}x^2 + \ldots$

Using **7**: compare $(1 + x)^n$ with $(1 - 3x)^7$

$= 1 + 7(-3x) + \dfrac{7 \times 6 \times (-3x)^2}{2!} + \ldots$

Replace n by 7 and x by $-3x$

$= 1 - 21x + 189x^2 + \ldots$

Simplify

(b) $(2 + 5x)^5 = \left(2\left(1 + \dfrac{5x}{2}\right)\right)^5$

Using **8**: take out a common factor of 2

$= 2^5\left(1 + \dfrac{5x}{2}\right)^5$

$2^5 = 32$

$= 32\left(1 + 5 \times \left(\dfrac{5x}{2}\right) + \dfrac{5 \times 4}{2!}\left(\dfrac{5x}{2}\right)^2 + \ldots\right)$

Replace n by 5 and x by $\dfrac{5x}{2}$ in the expansion of $(1 + x)^n$

$= 32\left(1 + \dfrac{25x}{2} + \dfrac{125x^2}{2} + \ldots\right)$

$= 32 + 400x + 2000x^2 + \ldots$

Multiply by 32

Worked exam style question 1

In the binomial expansion of $(2 + kx)^6$, where k is a constant, the coefficient of x^2 is 60. Calculate:

(a) the value of k, **(b)** the value of the coefficient of x^3 in the expansion.

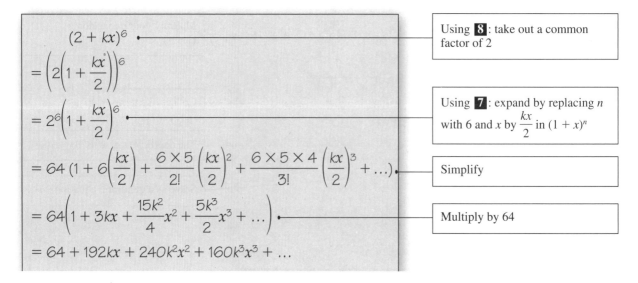

$(2 + kx)^6$

Using **8**: take out a common factor of 2

$= \left(2\left(1 + \dfrac{kx}{2}\right)\right)^6$

$= 2^6\left(1 + \dfrac{kx}{2}\right)^6$

Using **7**: expand by replacing n with 6 and x by $\dfrac{kx}{2}$ in $(1 + x)^n$

$= 64\left(1 + 6\left(\dfrac{kx}{2}\right) + \dfrac{6 \times 5}{2!}\left(\dfrac{kx}{2}\right)^2 + \dfrac{6 \times 5 \times 4}{3!}\left(\dfrac{kx}{2}\right)^3 + \ldots\right)$

Simplify

$= 64\left(1 + 3kx + \dfrac{15k^2}{4}x^2 + \dfrac{5k^3}{2}x^3 + \ldots\right)$

Multiply by 64

$= 64 + 192kx + 240k^2x^2 + 160k^3x^3 + \ldots$

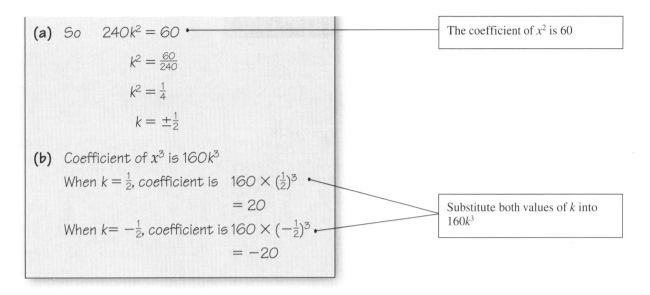

(a) So $\quad 240k^2 = 60$ ⟵ The coefficient of x^2 is 60

$$k^2 = \frac{60}{240}$$

$$k^2 = \frac{1}{4}$$

$$k = \pm\frac{1}{2}$$

(b) Coefficient of x^3 is $160k^3$

When $k = \frac{1}{2}$, coefficient is $\quad 160 \times (\frac{1}{2})^3$

$$= 20$$

When $k = -\frac{1}{2}$, coefficient is $160 \times (-\frac{1}{2})^3$

$$= -20$$

Substitute both values of k into $160k^3$

Worked exam style question 2

(a) Expand $(1 + 3x)^8$ in ascending powers of x up to and including the term in x^3.

(b) By substituting a suitable value of x into part **(a)**, calculate an approximate value to 1.03^8.

(a) $(1 + 3x)^8 = 1 + 8(3x) + \dfrac{8 \times 7}{2!}(3x)^2$

Using **7**: replace n by 8 and x by $3x$ in the expansion of $(1 + x)^n$

$$+ \dfrac{8 \times 7 \times 6}{3!}(3x)^3 + \ldots$$

$$= 1 + 24x + 252x^2 + 1512x^3 + \ldots$$

(b) Compare $(1 + 3x)^8$ with $(1.03)^8$ ⟵ Calculate the value of x

So $\quad 3x = 0.03$

$$x = 0.01$$

$$(1.03)^8 \approx 1 + 24(0.01) + 252(0.01)^2 + 1512(0.01)^3$$

Substitute $x = 0.01$ into both sides of your expansion

$$= 1 + 0.24 + 0.0252 + 0.001\,512$$

$$= 1.266\,712$$

So $(1.03)^8 \approx 1.266\,712$

Revision exercise 5

1 Write down the expansions of:
  (a) $(3x - 2)^4$ (b) $(2x + y)^5$.

2 Fully expand the expression $(1 - x)(1 + 2x)^4$.

3 The coefficient of x^2 in the expansion of $(2 + ax)^4$ is 48.
  Find possible values of the constant a.

4 Use the binomial expansion to find the first three terms in the expansions of:
  (a) $(1 - 2x)^6$ (b) $(2 + x)^5$.

5 Given that
$$(2 + x)^{10} \equiv A + Bx + Cx^2 + \dots$$
  find the values of the integers A, B and C.

6 Write down the first four terms in the expansion of $(1 + 3x)^6$.
  By substituting an appropriate value of x (which should be stated) find an approximate value of 1.03^6.

7 (a) Given that
$$(3 + x)^5 + (3 - x)^5 = A + Bx^2 + Cx^4$$
  find the value of the constants A, B and C.
  (b) Using the substitution $y = x^2$ and your answer to part (a) solve
$$(3 + x)^5 + (3 - x)^5 = 1686.$$

8 (a) Find the coefficient of x^5 and x^6 in the binomial expansion of $(2 + x)^9$.
  (b) The coefficient of x^5 and x^6 in the binomial expansion of $(2 + kx)^9$ are equal. Find the value of the constant k.

Test yourself	What to review
	If your answer is incorrect
1 Expand $\left(x + \dfrac{1}{x}\right)^4$, simplifying your coefficients.	*Review Heinemann Book C2 pages 70–71* *Revise for C2 page 29* *Example 1*
2 When $(1 - 3x)^p$ is expanded the coefficient of x^2 is 54. Given that $p > 0$, use this information to find: (a) the value of the constant p, (b) the coefficient of x.	*Review Heinemann Book C2 pages 75–76* *Revise for C2 page 29* *Worked exam style question 1*

3 Find the first four terms in the binomial expansion of $(3 - x)^6$.

Review Heinemann Book C2 page 76
Revise for C2 page 30
Example 3

4 (a) Use the binomial series to expand $(2 - x)^{10}$ in ascending powers of x up to and including the term in x^3, giving each coefficient as an integer.

(b) Use your series expansion, with a suitable value of x, to obtain an estimate for 1.98^{10}, giving your answer to 2 decimal places.

Review Heinemann Book C2 pages 74–76
Revise for C2 page 31
Worked exam style question 2

Test yourself answers

1 $x^4 + 4x^2 + 6 + \dfrac{4}{x^2} + \dfrac{1}{x^4}$ **2 (a)** $p = 4$ **(b)** -12 **3** $729 - 1458x + 1215x^2 - 540x^3$ **4 (a)** $1024 - 5120x + 11\,520x^2 - 15\,360x^3$ **(b)** 926.09

Radian measure and its applications

6

Key points to remember

1 A radian is the angle subtended at the centre of a circle by an arc whose length is equal to that of the radius of the circle.

If the arc AB of a circle, with centre O and radius r, has length r, then $\angle AOB$ is 1 radian (1^c or 1 rad).

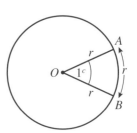

2 You can convert radians to degrees and vice-versa, using the relationship π rad $= 180°$.

$$1 \text{ rad} = \frac{180°}{\pi}, \; 1° = \frac{\pi}{180} \text{ rad}$$

3 The arc length l subtending an angle θ at the centre of a circle of radius r is $l = r\theta$.

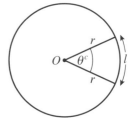

4 The area A of a sector of a circle of radius r is $A = \frac{1}{2}r^2\theta$.

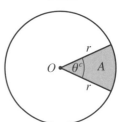

5 The area A of a segment of a circle of radius r is $A = \frac{1}{2}r^2(\theta - \sin\theta)$.

[In **3**, **4** and **5** θ must be in radians]

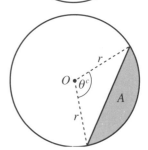

Example 1

Convert: **(a)** $125°$ to radians, giving your answer in terms of π,

(b) $\dfrac{\pi}{24}$ radians to degrees.

(a) $125° = 125 \times \dfrac{\pi}{180}$ rad

$\qquad = \dfrac{25}{36}\pi$

> Using **2**: $1° = \dfrac{\pi}{180}$ rad

(b) $\dfrac{\pi}{24}$ rad $= \dfrac{\pi}{24} \times \dfrac{180°}{\pi}$

$\qquad = 7\tfrac{1}{2}°$

> Using **2**: 1 rad $= \dfrac{180°}{\pi}$

Example 2

In the diagram AOB is a circular sector
of radius 12 cm with $\angle AOB = 0.65$ radians.
Calculate:

(a) the arc length AB,

(b) the area of the sector.

(a) Arc length $= 12 \times 0.65$ cm

$\qquad = 7.8$ cm

> Using **3**: with $r = 12$ and $\theta = 0.65$

(b) Area of sector $= \tfrac{1}{2} \times (12)^2 \times 0.65$ cm^2

$\qquad = 46.8$ cm^2

> Using **4**

Example 3

The area of a sector AOB, of a circle of radius 5 cm, is 16 cm^2.
Given that AB subtends an angle of θ radians at the centre O of the
circle, calculate the value of θ.

$16 = \tfrac{1}{2}(25)\theta$

So $\qquad \theta = \dfrac{16}{12.5} = 1.28$

> Using **4**: with $A = 16$ and $r = 5$

Example 4

In the diagram, the minor arc AB of the circle, with centre O and radius 15 cm, subtends an angle of $\dfrac{\pi}{5}$ radians at O, as shown.

Calculate, giving your answer in terms of π, the area of the major sector AOB.

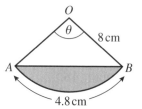

The shaded region is the major segment.

The angle subtended by the major sector AOB at O is the

reflex angle $\left(2\pi - \dfrac{\pi}{5}\right) = \dfrac{9\pi}{5}$

Area of (major) sector $= \frac{1}{2}(15)^2\left(\dfrac{9\pi}{5}\right)$ cm^2

$= 202.5\pi$ cm^2

The total angle at $O = 2\pi$ rad

Using **4**: with $r = 15$ and $\theta = \dfrac{9\pi}{5}$

Leave answer in terms of π

Worked exam style question 1

In the diagram the arc AB of a circle, with centre O and radius 8 cm, has length 4.8 cm, as shown. Given that $\angle AOB = \theta$ radians, calculate:

(a) the value of θ,

(b) the area of the sector AOB,

(c) the area of the shaded segment, giving your answer to 3 significant figures.

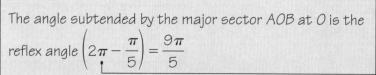

(a) $4.8 = 8\theta$

So $\theta = 0.6$

(b)   Area of sector $= \frac{1}{2}(8)^2(0.6)$ cm^2

$= 19.2$ cm^2

(b)   Area of segment $= \frac{1}{2}(8)^2(0.6 - \sin 0.6^c)$

$= 1.13$ cm^2 (3 s.f.)

Using **3**: with $l = 4.8$ and $r = 8$

Using **4**

Using **5**: remember to use radian mode

Worked exam style question 2

The triangle ABC has $AB = 9$ cm, $BC = 13$ cm and $CA = 10$ cm, and $\angle BAC = \theta$ radians. A circle centre A and radius 4.5 cm intersects AB and AC at D and E respectively, as shown.

(a) Show that $\theta = 1.504$, to 3 decimal places.

(b) Calculate:
   (i) the perimeter of the shaded region $BCED$,
   (ii) the area of the shaded region $BCED$
   giving your answers to 3 significant figures.

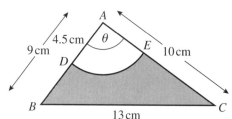

(a) Using the cosine rule in $\triangle ABC$,

$$\cos \theta = \frac{9^2 + 10^2 - 13^2}{2(9)(10)} = 0.0\dot{6}$$

$$\Rightarrow \quad \theta = \cos^{-1}(0.0\dot{6}) = 1.504 \ (3 \text{ d.p.})$$

> As the lengths of three sides of the triangle are given, use cosine rule: $\cos A = \dfrac{b^2 + c^2 - a^2}{2bc}$

(b) (i) Perimeter $= DB + BC + CE + \text{arc } DE$

$$= [(9 - 4.5) + 13 + (10 - 4.5) + 4.5\theta] \text{ cm}$$

$$= 29.768... \text{ cm}$$

Perimeter $= 29.8$ cm (3 s.f.)

> Using **3**

(ii) Area of shaded region
$= \text{area of } \triangle ABC - \text{area of sector } ADE$

$$= \tfrac{1}{2}(9)(10) \sin \theta - \tfrac{1}{2}(4.5)^2(\theta)$$

$$= 29.7 \text{ cm}^2 \ (3 \text{ s.f.})$$

> Using area $= \frac{1}{2}bc \sin A$ in $\triangle ABC$

> Using **4**

Worked exam style question 3

In the diagram, the shaded region represents the horizontal upper surface s of a concrete step. The arc AB is part of a circle, centre O and radius r m. Given that angle $AOB = 0.9$ radians, and that the perimeter of s is 12.4 m:

(a) show that $r = 7.0$, to 2 significant figures,

(b) calculate the area of s, giving your answer to 2 significant figures.

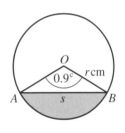

(a) The perimeter of the surface consists of chord AB and arc AB.

In $\triangle AOM$, where M is the mid-point of AB,

$$\frac{\tfrac{1}{2}AB}{r} = \sin 0.45^c$$

So chord $AB = 2r \sin 0.45^c$

As the perimeter $= 12.4$ m,

$$0.9r + 2r \sin 0.45^c = 12.4$$

$$\Rightarrow r(0.9 + 2 \sin 0.45^c) = 12.4$$

so

$$r = \frac{12.4}{0.9 + 2 \sin 0.45^c}$$

$$= 7.0059... = 7.0 \ (2 \text{ s.f.})$$

> As $\triangle AOB$ is isosceles, use the symmetry property and right-angled triangle trigonometry

> Arc AB + chord AB = 12.4 m

> Using **3**

> Factorise

> Use calculator in radian mode

(b) Area of segment $= \frac{1}{2}r^2 (0.9 - \sin 0.9^c)$ ← Using **5**

$= 2.863... \, m^2$

So area of s $= 2.9 \, m^2$ (2 s.f.)

This answer uses the stored calculator value for r

Revision exercise 6

1 (a) Convert the following angles, given in radians, to degrees:

    (i) $\dfrac{2\pi}{3}$ **(ii)** $\dfrac{\pi}{400}$ **(iii)** $\dfrac{7\pi}{9}$ **(iv)** $\dfrac{5\pi}{144}$.

(b) Convert the following angles to radians, giving your answer as $k\pi$, where k is a fraction in its simplest form.

    (i) $6°$ **(ii)** $13.5°$ **(iii)** $105°$ **(iv)** $420°$

2 Referring to the diagram of a circular sector with radius r cm and angle θ radians, calculate when:

    (i) $r = 2.6$, $\theta = 0.9$,

    (ii) $r = 17.5$, $\theta = \dfrac{2\pi}{7}$,

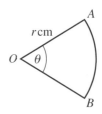

(a) the length of arc AB,

(b) the area of the sector AOB.

[Leave your answers to **(ii)** in term of π]

3 A sector of a circle, centre O and radius 8 cm, contains an angle of $\dfrac{\pi}{4}$ radians at O.

Find, in terms of π:

(a) the length of the arc of the sector,

(b) the area of the sector.

4 Calculate the area of the shaded segment in the diagram shown, giving your answer to 3 significant figures.

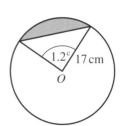

5 The diagram shows the sector of a circle of radius 16.5 cm. The angle contained by the sector is 0.7 radians. Calculate the perimeter of the sector.

6 The arc length of a sector of a circle, of radius 12 cm and centre O, is 7.68 cm. Calculate, in radians, the angle of the sector.

7 The area of a circular sector of radius 5 cm is 20 cm².
Calculate the length of the arc of the sector.

8 Find the area of the shaded sector in each of the following circles
with centre *C*. Leave your answer in terms of π, where appropriate.

(a)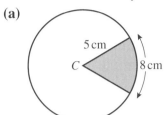

(b)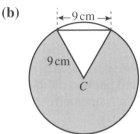

9 The sector *AOB* of a circle, centre *O* and radius *r* cm, has area
28.8 cm² and contains an angle of 1.2 radians at *O*.
Prove that $r = 4\sqrt{3}$.

10 A brooch has the shape of a sector of a circle of radius 4 cm.
The perimeter of the brooch is 11 cm.
Calculate the area of the brooch.

11 A sector of circle has radius 20 cm and contains an angle of 54°,
as shown.

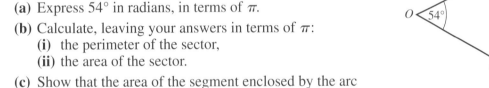

(a) Express 54° in radians, in terms of π.

(b) Calculate, leaving your answers in terms of π:
  (i) the perimeter of the sector,
  (ii) the area of the sector.

(c) Show that the area of the segment enclosed by the arc
    AB and the chord *AB* is approximately 26.7 cm².

12 In the diagram *AB* is the diameter of a circle, with centre *O* and
radius 8 cm, and $\angle BOC = \theta$ radians. The shaded segment has
area *A* cm² and the area of triangle *BOC* = *B* cm². Given that
$A = 2B$, show that $\pi = \theta + 3 \sin \theta$.

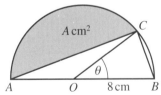

13 In the diagram, *AB* is the arc of a circle, centre *O* and radius
10 cm. The points *C* and *D* are such that *OC* = 5 cm,
OD = 4 cm. Angle *AOB* = 0.75 radians. Calculate, giving
your answer to 3 significant figures:

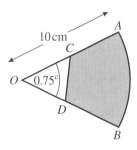

(a) the area of the shaded region,

(b) the perimeter of the shaded region.

14 The diagrams below show the cross-sections of two biscuits.

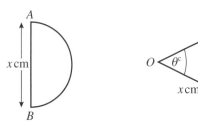

One is semi-circular with diameter $AB = x$ cm, the other is a circular sector DOC, with radius x cm and angle θ radians, as shown. The two cross-sections have the same perimeter.

(a) Show that $\theta = \dfrac{\pi}{2} - 1$.

(b) Given that $x = 6$, find in terms of π, the difference in the areas of the shapes.

15 A triangular piece of ground has sides AB, BC and AC of lengths 70 m, 90 m and 120 m respectively. A circular arc, centre A and radius 70 m, joining B to the point D, on AC, is marked out on the ground. A lawn is to be laid in the sector ABD.

(a) Show that $\angle BAD = 0.841$ radians, to 3 significant figures.

(b) Calculate the area of the ground which will not be lawn, giving your answer to 3 significant figures.

16 A sector of a circle, of radius r cm, contains an angle of 0.8 radians at the centre of the circle. The sector has perimeter P cm and area A cm^2, where $A + P = 31.2$.

(a) Show that r satisfies the equation $r^2 + 7r - 78 = 0$.

(b) Calculate the values of A and P.

17 In the diagram AB and DC are arcs of circles with centre O, such that $OA = OB = r$ cm, $AC = BD = 5$ cm and $\angle DOC = 0.48$ radians.
Given that the area of the shaded region $ABCD = 54$ cm^2, calculate the value of r.

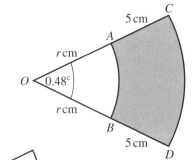

18 A thin metal plate has the shape of a circular sector, with radius r cm and angle θ radians, as shown in the diagram. The perimeter of the plate is 36 cm.

(a) Show that $\theta = \dfrac{36 - 2r}{r}$.

(b) Given that the area of the plate is A cm^2, prove that $A = 81 - (r - 9)^2$.

(c) Find the values of r and θ which would give a plate with perimeter 36 cm^2 and a maximum area.

Hint: See Book C1, Section 2.3, for method of completing the square.

Test yourself	**What to review**

If your answer is incorrect

1 (a) Convert: **(i)** $\dfrac{2\pi}{5}$ **(ii)** 1.5 radians to degrees.

   (b) Convert 25° to radians, leaving your answer in terms of π.

Review Heinemann Book C2 pages 80–81
Revise for C2 page 35
Example 1

2 Referring to the diagram shown, calculate:

   (a) the arc length AB,

   (b) the area of the sector AOB.

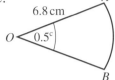

Review Heinemann Book C2 pages 82–86
Revise for C2 page 35
Example 2

3 The area of the sector of a circle of radius 15 cm is 240 cm^2. Find the perimeter of the sector.

Review Heinemann Book C2 page 85
Revise for C2 page 35
Example 3

4 The minor sector AOB in the circle of radius 6 cm contains an angle of $\dfrac{\pi}{18}$ at the centre, as shown. Find the area of the major sector AOB, giving your answer in terms of π.

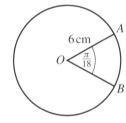

Review Heinemann Book C2 page 86
Revise for C2 page 36
Example 4

5 The chord AB of a circle, with centre O and radius 11.5 cm, has length 13 cm and subtends an angle of θ radians at O, as shown.

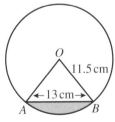

   (a) Show that $\theta = 1.20$, to 3 significant figures.

   (b) Calculate, to 3 significant figures, the area of the shaded segment.

Review Heinemann Book C2 page 85
Revise for C2 page 36
Worked exam style question 1, page 37
Worked exam style question 3

Test yourself answers

1 **(a) (i)** 72° **(ii)** 85.9° (nearest 0.1°) **(b)** $\dfrac{5\pi}{36}$ 2 **(a)** 3.4 cm **(b)** 11.56 cm^2 3 62 cm   4 35π cm^2 5 **(b)** 17.8 cm^2

Geometric sequences and series

7

Key points to remember

1   In a geometric series you get from one term to the next by multiplying by a constant called the common ratio.

2   The formula for the nth term $= ar^{n-1}$ where $a =$ first term and $r =$ common ratio.

3   The formula for the sum to n terms is
$$S_n = \frac{a(1 - r^n)}{1 - r} \quad or \quad S_n = \frac{a(r^n - 1)}{r - 1}.$$

4   The sum to infinity exists if $|r| < 1$ and is $S_\infty = \dfrac{a}{1 - r}$.

Example 1

The second term of a geometric series is 8 and the 5th term is -1. Find in any order:

(a) the common ratio,   **(b)** the first term,   **(c)** the tenth term.

(a)   Let $a =$ first term and $r =$ common ratio

2nd term $= ar = 8$ ①

5th term $= ar^4 = -1$ ②

Using **2**: nth term $= ar^{n-1}$
with $n = 2$
and $n = 5$

② ÷ ① $\dfrac{\cancel{a}r^4}{\cancel{a}r} = -\frac{1}{8}$

$r^3 = -\frac{1}{8}$

$r = -\frac{1}{2}$

(b)   Common ratio is $-\frac{1}{2}$.

Substitute $r = -\frac{1}{2}$ in ①

$a \times (-\frac{1}{2}) = 8$

Use $ar = 8$ with $r = -\frac{1}{2}$

$a = 8 \times (-2)$

$a = -16$

First term is -16.

(c) $\quad$ 10th term $= ar^9$

$\qquad = (-16) \times (-\tfrac{1}{2})^9$

$\qquad = \dfrac{-2^4}{-2^9}$

$\qquad = \dfrac{1}{2^5}$

$\qquad = \dfrac{1}{32}$

Tenth term is $\tfrac{1}{32}$

> Using **2**: 10th term $= ar^9$ with $a = -16$ and $r = -\tfrac{1}{2}$

Example 2

I invest £4000 in a bank which is offering interest at a rate of 3.1% per annum.
How long will it be before I double my money if the interest rate stays constant?

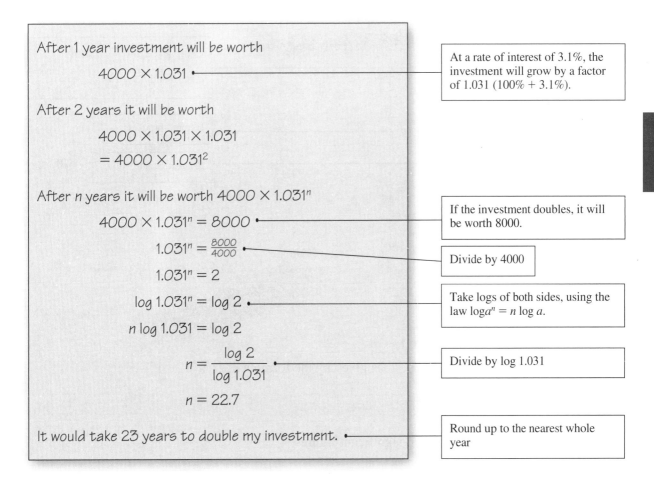

After 1 year investment will be worth

$\qquad 4000 \times 1.031$

> At a rate of interest of 3.1%, the investment will grow by a factor of 1.031 (100% + 3.1%).

After 2 years it will be worth

$\qquad 4000 \times 1.031 \times 1.031$

$\qquad = 4000 \times 1.031^2$

After n years it will be worth 4000×1.031^n

$\qquad 4000 \times 1.031^n = 8000$

> If the investment doubles, it will be worth 8000.

$\qquad 1.031^n = \dfrac{8000}{4000}$

> Divide by 4000

$\qquad 1.031^n = 2$

$\qquad \log 1.031^n = \log 2$

> Take logs of both sides, using the law $\log a^n = n \log a$.

$\qquad n \log 1.031 = \log 2$

$\qquad n = \dfrac{\log 2}{\log 1.031}$

> Divide by log 1.031

$\qquad n = 22.7$

It would take 23 years to double my investment.

> Round up to the nearest whole year

Example 3

Calculate $\displaystyle\sum_{r=1}^{10} 3 \times 5^r$.

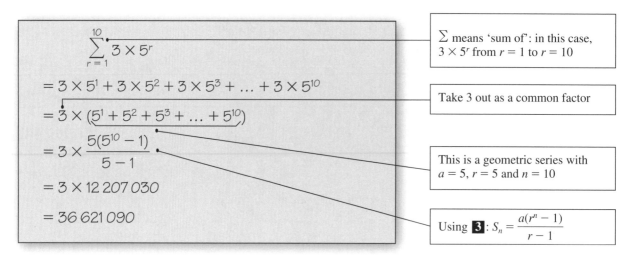

$$\sum_{r=1}^{10} 3 \times 5^r$$

Σ means 'sum of': in this case, 3×5^r from $r = 1$ to $r = 10$

$$= 3 \times 5^1 + 3 \times 5^2 + 3 \times 5^3 + \ldots + 3 \times 5^{10}$$

$$= 3 \times (5^1 + 5^2 + 5^3 + \ldots + 5^{10})$$

Take 3 out as a common factor

$$= 3 \times \frac{5(5^{10} - 1)}{5 - 1}$$

This is a geometric series with $a = 5$, $r = 5$ and $n = 10$

$$= 3 \times 12\,207\,030$$

$$= 36\,621\,090$$

Using **3**: $S_n = \dfrac{a(r^n - 1)}{r - 1}$

Example 4

Find the common ratio of a geometric series with first term 10 and sum to infinity of 25.

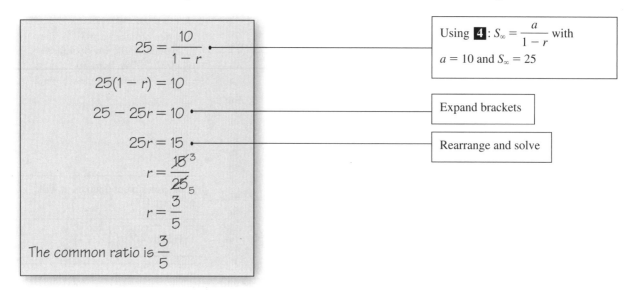

$$25 = \frac{10}{1 - r}$$

Using **4**: $S_\infty = \dfrac{a}{1 - r}$ with $a = 10$ and $S_\infty = 25$

$$25(1 - r) = 10$$

$$25 - 25r = 10$$

Expand brackets

$$25r = 15$$

Rearrange and solve

$$r = \frac{\cancel{15}^{\,3}}{\cancel{25}_{\,5}}$$

$$r = \frac{3}{5}$$

The common ratio is $\dfrac{3}{5}$

Worked exam style question 1

A sequence of numbers $u_1, u_2, \ldots, u_n, \ldots$ is given by the formula $u_n = 5(\frac{4}{5})^n - 1$ where n is a positive integer.

(a) Find the values of u_1, u_2 and u_3.

(b) Show that $\displaystyle\sum_{n=1}^{20} u_n = -0.2306$ to 4 significant figures.

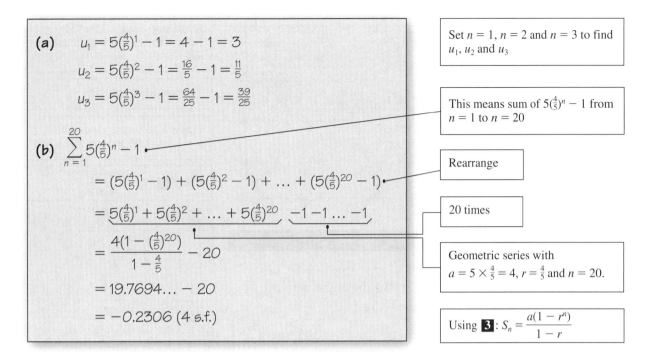

(a) $u_1 = 5(\frac{4}{5})^1 - 1 = 4 - 1 = 3$

$u_2 = 5(\frac{4}{5})^2 - 1 = \frac{16}{5} - 1 = \frac{11}{5}$

$u_3 = 5(\frac{4}{5})^3 - 1 = \frac{64}{25} - 1 = \frac{39}{25}$

Set $n = 1$, $n = 2$ and $n = 3$ to find u_1, u_2 and u_3

(b) $\sum_{n=1}^{20} 5(\frac{4}{5})^n - 1$

This means sum of $5(\frac{4}{5})^n - 1$ from $n = 1$ to $n = 20$

$= (5(\frac{4}{5})^1 - 1) + (5(\frac{4}{5})^2 - 1) + \ldots + (5(\frac{4}{5})^{20} - 1)$

Rearrange

$= \underbrace{5(\frac{4}{5})^1 + 5(\frac{4}{5})^2 + \ldots + 5(\frac{4}{5})^{20}}, \underbrace{-1 - 1 \ldots - 1}$

20 times

$= \dfrac{4(1 - (\frac{4}{5})^{20})}{1 - \frac{4}{5}} - 20$

Geometric series with $a = 5 \times \frac{4}{5} = 4$, $r = \frac{4}{5}$ and $n = 20$.

$= 19.7694\ldots - 20$

$= -0.2306$ (4 s.f.)

Using **3**: $S_n = \dfrac{a(1 - r^n)}{1 - r}$

Worked exam style question 2

A savings scheme pays 6% per annum compound interest.
A deposit of £500 is invested into this scheme at the start of each year.

(a) Show that at the start of the third year, after the annual deposit has been made, the amount in the scheme is £1591.80.

(b) Find the amount in the scheme at the start of the twentieth year, after the annual deposit has been made.

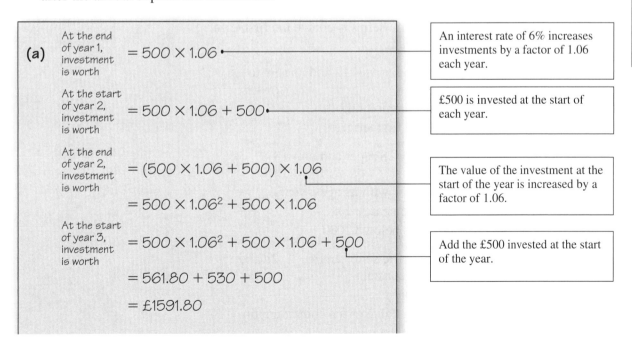

(a) At the end of year 1, investment is worth $= 500 \times 1.06$

An interest rate of 6% increases investments by a factor of 1.06 each year.

At the start of year 2, investment is worth $= 500 \times 1.06 + 500$

£500 is invested at the start of each year.

At the end of year 2, investment is worth $= (500 \times 1.06 + 500) \times 1.06$

$= 500 \times 1.06^2 + 500 \times 1.06$

The value of the investment at the start of the year is increased by a factor of 1.06.

At the start of year 3, investment is worth $= 500 \times 1.06^2 + 500 \times 1.06 + 500$

Add the £500 invested at the start of the year.

$= 561.80 + 530 + 500$

$= £1591.80$

(b) At the start of the 20th year, investment is worth

$$= 500 \times 1.06^{19} + 500 \times 1.06^{18} + \ldots + 500 \times 1.06 + 500$$

$$= 500 \left(1.06^{19} + 1.06^{18} + \ldots + 1.06 + 1\right)$$

$$= 500 \times \frac{1(1.06^{20} - 1)}{1.06 - 1}$$

$$= £18\,392.80$$

> At the start of year 3 the index is 2. So at the start of year 20 the index is 19.

> Take out a factor of 500

> This is a geometric series with $a = 1$, $r = 1.06$ and $n = 20$

> Using **3**: $S_n = \dfrac{a(r^n - 1)}{r - 1}$

Revision exercise 7

1 For the following geometric series $\quad 4 + 3.2 + 2.56 + \ldots$
Find: **(a)** the common ratio, **(b)** the eighth term,
(c) the sum to eight terms (correct to 5 s.f.).

2 A geometric series has a first term of 2 and a sixth term of $\frac{1}{16}$.
(a) Show that the common ratio is $\frac{1}{2}$.
(b) Find the sum to infinity of the series.
(c) Find, to 3 significant figures, the difference between the sum to infinity and the sum of the first ten terms.

3 The nth term of a sequence is u_n, where $u_n = 30 \times \left(\frac{3}{10}\right)^n$,
$n = 1, 2, 3, \ldots$
(a) Find the value of u_1 and u_2.
(b) Calculate $\displaystyle\sum_{n=1}^{15} u_n$ to three significant figures.

4 The sum to infinity of a geometric series is double the first term. Find the value of the common ratio.

5 The second term of a geometric series is 12 and its sum to infinity is 50. Find:
(a) all possible values of the common ratio,
(b) corresponding values of the first term.

6 The price of a car depreciates by 18% per annum.
If its new price is £16 000, find:
(a) a formula linking its value £V with its age a years,
(b) its value after 4 years (to the nearest £),
(c) the year when its value falls below £2000.

7 The first three terms of a geometric series are $p - 1$, $2p$ and $4p + 6$ respectively, where p is constant.
(a) Find the value of the constant p.
(b) Calculate the corresponding value of the common ratio.
(c) Find the sum to ten terms of the series.

8 The sum to infinity of a geometric series is three times the sum to five terms. Find the common ratio to 3 decimal places.

9 Ian is running the London Marathon for charity. He manages to run the first mile in 6 minutes but takes 5% longer for each subsequent mile of the 26 mile race.
  (a) Prove that he runs the tenth mile in 9 minutes 18 seconds (to the nearest second).
  (b) Find his time for the race.

10 There are 400 deer in a herd in January 2004.
  Their population grows at 10% per year but 30 deer are culled at the end of each year.
  (a) Show that in January 2006 there are 421 in the herd.
  (b) How many deer will be in the herd in January 2015?

Test yourself	What to review
	If your answer is incorrect
1 In the geometric series $10 + 2 + 0.4 + \dots$ calculate: **(a)** the common ratio, **(b)** the sum to infinity.	*Review Heinemann Book C2 pages 94 and 104* *Revise for C2 page 44 Example 4*
2 A geometric series has third term of 108 and fifth term of 27. Calculate: **(a)** the common ratio, given that it is positive, **(b)** the first term in the series, **(c)** the sum to ten terms in the series.	*Review Heinemann Book C2 pages 96 and 100* *Revise for C2 page 44 Example 1*
3 A sequence of numbers u_1, u_2, u_3 is given by the formula $u_n = 12\left(\frac{1}{2}\right)^n - 3$, where n is a positive integer. **(a)** Find the values of u_1, u_2 and u_3. **(b)** Show $\sum_{n=1}^{20} u_n = -48$ (6 s.f.)	*Review Heinemann Book C2 page 102* *Revise for C2 page 44 Worked exam style question 1*
4 A savings scheme pays 6% compound interest. Sarah invests £2000 at the start of each year. **(a)** Show that at the start of the third year there is £6367.20 in the scheme. **(b)** Find the amount in the scheme at the start of the twentieth year.	*Review Heinemann Book C2 page 101* *Revise for C2 page 45 Worked exam style question 2*

Test yourself answers

1 (a) 0.2 (b) 12.5   2 (a) $\frac{1}{2}$ (b) 432 (c) 863.15625   3 (a) 3, 0, −1.5   4 (b) £73 571.18

Graphs of trigonometric functions

8

Key points to remember

1 The x–y plane is divided into quadrants.
Angles are measured from the positive x-axis.
Anticlockwise angles are positive; clockwise
angles are negative.

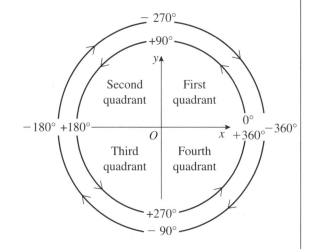

2 For all values of θ, the definitions of $\sin \theta$, $\cos \theta$ and
$\tan \theta$ are taken to be

$$\sin \theta = \frac{y}{r}, \cos \theta = \frac{x}{r}, \tan \theta = \frac{y}{x},$$

where x and y are the coordinates of P as the line
OP, of length r, makes an angle θ with the positive x-axis.

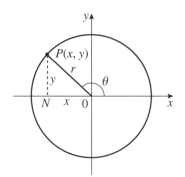

3 In the first quadrant, where θ is acute,
All trigonometric functions are positive.
In the second quadrant, where θ is obtuse,
only **S**ine is positive.
In the third quadrant, where θ is reflex,
$180° < \theta < 270°$, only **T**angent is positive.
In the fourth quadrant, where θ is reflex,
$270° < \theta < 360°$, only **C**osine is positive.

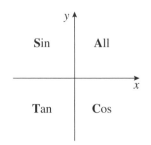

4 The trigonometric ratios of angles equally inclined to the
x-axis are related:
$\sin(180 - \theta)° = \sin \theta°$,
$\sin(180 + \theta)° = -\sin \theta°$,
$\sin(360 - \theta)° = -\sin \theta°$,
$\cos(180 - \theta)° = -\cos \theta°$,
$\cos(180 + \theta)° = -\cos \theta°$,
$\cos(360 - \theta)° = \cos \theta°$,
$\tan(180 - \theta)° = -\tan \theta°$,
$\tan(180 + \theta)° = \tan \theta°$,
$\tan(360 - \theta)° = -\tan \theta°$.

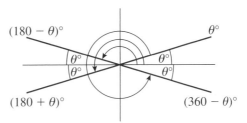

5 The trigonometric ratios of 30°, 45° and 60° have exact
forms, given below:

$$\sin 30° = \frac{1}{2}, \qquad \cos 30° = \frac{\sqrt{3}}{2}, \qquad \tan 30° = \frac{1}{\sqrt{3}} = \frac{\sqrt{3}}{3}.$$

$$\sin 45° = \frac{1}{\sqrt{2}} = \frac{\sqrt{2}}{2}, \qquad \cos 45° = \frac{1}{\sqrt{2}} = \frac{\sqrt{2}}{2}, \qquad \tan 45° = 1.$$

$$\sin 60° = \frac{\sqrt{3}}{2}, \qquad \cos 60° = \frac{1}{2}, \qquad \tan 60° = \sqrt{3}.$$

6 The graph of $y = \sin \theta°$ is
shown opposite:

The period (smallest interval in
which the function repeats
itself) of the sine function is
360° (or 2π for $y = \sin \theta$,
where θ is measured in
radians).
Periodic properties are
$\sin(\theta \pm 360)° = \sin \theta°$.

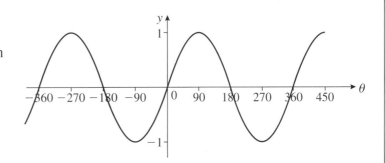

7 The graph of $y = \cos \theta°$ is
shown opposite:

The period of the cosine
function is 360° (or 2π for
$y = \cos \theta$, where θ is
measured in radians).
Periodic properties are
$\cos(\theta \pm 360)° = \cos \theta°$.

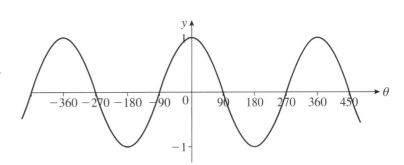

7 The graph of $y = \tan \theta°$ is shown below:

Asymptotes occur at $\theta = (2n + 1)90°$, $n \in Z$.

The tangent function has a period of 180°, (or π radians). Periodic properties are tan $(\theta \pm 180)° = \tan \theta°$.

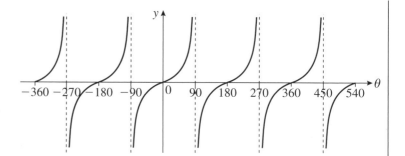

8 Some other useful properties are:

$\sin(-\theta) = -\sin\theta$, $\tan(-\theta) = -\tan\theta$, $\cos(-\theta) = \cos\theta$

(Sine and tangent are examples of odd functions; cosine is an example of an even function.)

$\sin(90 - \theta)° = \cos\theta°$, $\cos(90 - \theta)° = \sin\theta°$.

Example 1

Find the exact values of: **(a)** $\sin 135°$ **(b)** $\sin\dfrac{7\pi}{6}$.

(a) $\sin 135° = \sin(180 - 45)° = \sin 45°$

$= \dfrac{\sqrt{2}}{2}$

(b) $\sin\dfrac{7\pi}{6} = \sin(\pi + \dfrac{\pi}{6}) = -\sin\dfrac{\pi}{6}$

$= -\dfrac{1}{2}$

Using **4**: sine is positive in 2nd quadrant, $\sin(180 - \theta) = \sin\theta$

Using **5**

Using **4**: sine is negative in 3rd quadrant, so $\sin(\pi + \theta) = -\sin\theta$

Using **5** and $\dfrac{\pi}{6}$ rad $= 30°$

Example 2

Express: **(a)** $\cos(-100)°$ in terms of $\cos y°$, where $y°$ is an acute angle,

(b) $\sin(\theta - 180)°$ in terms of $\sin\theta°$, where $\theta°$ is an acute angle.

(a)

(S)   (A)

80°   −100°

(T)   (C)

So $\cos(-100)° = -\cos 80°$

Using **4**: $-100°$ is in the 3rd quadrant, where cos is negative, and the acute angle to the horizontal is 80°

Using **9**: $\cos(-\theta)° = \cos\theta°$, so $\cos(-100)° = \cos 100°$

Using **4**: $\cos(180 - \theta)° = -\cos\theta°$, with $\theta = 80$

(b)

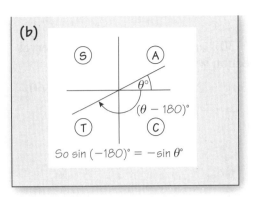

So $\sin(-180)° = -\sin\theta°$

Using **4** : $(\theta - 180)°$ is in the 3rd quadrant, where sine is negative, and the acute angle to the horizontal is $\theta°$

Using **9** : $\sin(\theta - 180)° = -\sin(180 - \theta)°$

Using **4** : $\sin(180 - \theta)° = \sin\theta°$

Example 3

Write down the coordinates, in the interval $-180 < x < 180$, of the maximum and minimum points on the curves with the following equations:

(a) $y = 3\sin x°$ **(b)** $y = \cos 2x°$.

(a) Maximum is at $(90, 3)$, minimum is at $(-90, -3)$

(b) Maximum is at $(0, 1)$, minima are at $(-90, -1)$ and $(90, -1)$

Using **6** and remembering that $y = 3\sin x°$ is a vertical stretch of $y = \sin x°$ by scale factor 3. (See Book C1, Section 4.6)

Using **7** and remembering that $y = \cos 2x°$ is a horizontal stretch of $y = \cos x°$ by scale factor $\frac{1}{2}$. (See Book C1, Section 4.6)

Example 4

Write down the period of each of the following functions:

(a) $\sin 5x°$ **(b)** $\tan\left(\dfrac{x}{3}\right)°$

(c) $\tan\left(\theta + \dfrac{\pi}{6}\right)$.

Using **6** : the graph of $y = \sin x°$ is stretched horizontally with scale factor $\frac{1}{5}$ to produce $y = \sin 5x°$, so a complete wave only has a width of $\frac{1}{5} \times 360°$. In general, the period of $\sin kx°$ is $\dfrac{360°}{k}$.

(a) Period of $\sin x°$ is $360°$, so period of $\sin 5x°$ is $72°$

(b) Period of $\tan x°$ is $180°$, so period of $\tan\left(\dfrac{x}{3}\right)°$ is $540°$

(c) Period of $\tan\left(\theta + \dfrac{\pi}{6}\right)$ is π

Using **8** : in general, the period of $\tan kx°$ is $\dfrac{180°}{k}$; here $k = \frac{1}{3}$

As the graph of $\tan\left(\theta + \dfrac{\pi}{6}\right)$ is a horizontal translation of that of $\tan\theta$, they have the same period. Note that the answer is in radians.

Worked exam style question 1

(a) Sketch, on the same set of axes, the graphs of $y = \sin(x + 30)°$ and $y = 3\cos x°$, in the interval $0 \leqslant x < 360$. Show the coordinates of all points of intersection with the axes.

(b) Deduce the number of solutions of the equation $\sin(x + 30)° - 3\cos x° = 0$, for $0 \leqslant x < 360$.

(a)

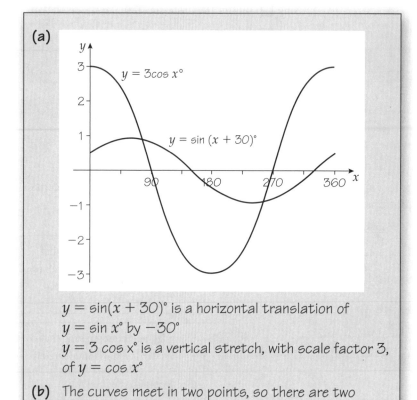

$y = \sin(x + 30)°$ is a horizontal translation of $y = \sin x°$ by $-30°$
$y = 3\cos x°$ is a vertical stretch, with scale factor 3, of $y = \cos x°$

(b) The curves meet in two points, so there are two solutions to the given equation.

Worked exam style question 2

The diagram shows part of the graph with equation $y = f(x)$. It crosses the x-axis at $A(110, 0)$ and B; it crosses the y-axis at $C(0, p)$. The point D is a minimum point.

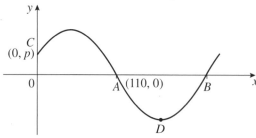

Given that $f(x) = \cos(x - q)°$, where $q > 0$, write down:

(a) the coordinates of **(i)** B **(ii)** D,
(b) the smallest value of q,
(c) the value of p, to 3 significant figures.

(a) (i) $(290, 0)$
    (ii) $(200, 0)$

Using **7**: x-coordinate of $B = 110° + 180°$ since
$y = \cos(x - q)°$ is a horizontal translation of $y = \cos x°$

Using **7**: add $90°$ to $110°$

(b) $q = 20$

$y = \cos(x - q)°$ is a horizontal translation of $y = \cos x°$
by $q°$ to the right. The first intersection of $y = \cos x°$
with the positive x-axis is at $90°$, so $q = 20$.
(Other values of q are $20 + 360n$, $n \in Z^+$)

(c) $p = \cos(-20)° = 0.940$ (3 s.f.)

Curve meets y-axis where $x = 0$,
so y-coordinate of C is $\cos(0 - 20)°$

Revision exercise 8

1 Use the appropriate graphs to write down the values of:

(a) $\sin 90°$
(b) $\sin(-180)°$
(c) $\sin 270°$
(d) $\sin 450°$
(e) $\cos(-90)°$
(f) $\cos 0°$
(g) $\cos 180°$
(h) $\cos 720°$
(i) $\tan 180°$
(j) $\tan(-180)°$
(k) $\sin \dfrac{3\pi}{2}$
(l) $\cos(-\pi)$.

2 Express:

(a) $\sin 280°$ in terms of $\sin x°$, where $x°$ is an acute angle,
(b) $\cos(-130)°$ in terms of $\cos y°$, where $y°$ is an acute angle,
(c) $\tan 185°$ in terms of $\tan z°$, where $z°$ is an acute angle,
(d) $\sin \dfrac{11\pi}{6}$ in terms of $\sin k\pi$, where $k < \tfrac{1}{2}$.

3 Given that $\sin x° = -\sin 50°$, $0 < x < 360$, use the quadrant
diagram, **4**, to find values of x.

4 Find the exact values of:

(a) $\sin 210°$
(b) $\sin(-225)°$
(c) $\cos 330°$
(d) $\tan 135°$
(e) $\cos\left(\dfrac{4\pi}{3}\right)$
(f) $\tan \dfrac{7\pi}{6}$.

5 Write down the periods of the following functions:

(a) $\sin 4\theta°$
(b) $3\cos\theta°$
(c) $\tan\left(\dfrac{\theta}{5}\right)°$
(d) $4 + \sin\theta$ (θ in radians).

6 Write down the equation, in the intervals given, of each of the asymptotes for the curve with equation:

(a) $y = \tan x°$, for $-300 < x < 150$

(b) $y = \tan (x + 45)°$, for $0 < x < 450$

(c) $y = \tan 3\theta$, for $-\dfrac{\pi}{3} < \theta < \pi$.

7 (a) Sketch the graph of $y = \sin x°$, in the interval $-180 \leqslant x \leqslant 360$.

(b) One point of intersection of the line $y = k$, $-1 < k < 0$, with $y = \sin x°$ has x-coordinate p, where $-90 < p < 0$.

Write down, in terms of p, the x-coordinates of other points of intersection in the given interval.

8 The diagram shows part of the curve with equation $y = -\sin x°$.

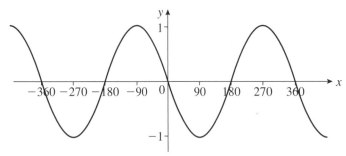

By considering appropriate transformations, state which of the following equations represent the same curve:

(a) $y = \cos (x - 90)°$ **(b)** $y = \cos (x + 90)°$

(c) $y = \sin (x - 180)°$ **(d)** $y = \sin (x + 180)°$

(e) $y = \sin (-x)°$ **(f)** $y = \sin (180 - x)°$.

9 Write down the equation of the curve resulting from each of the following transformations.

(a) A horizontal translation of $y = \sin (x + 30)°$ by 60 in the positive x-direction.

(b) A vertical translation of $y = \sin (x + 30)°$ by 3 in the positive y-direction.

(c) A horizontal stretch, with scale factor 3, of $y = \tan \theta$.

10 The diagram shows part of the curve C with equation $y = k \cos x°$, where k is a constant.

Write down:

(a) the values of: **(i)** k **(ii)** p,

(b) the equation of the curve resulting from stretching C horizontally by a scale factor of $\frac{1}{2}$.

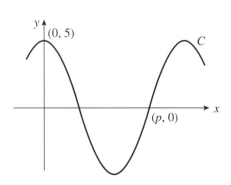

11 **(a)** Sketch the graph of $y = \sin\left(\frac{1}{2}\theta\right)$, in the interval $-3\pi \leqslant \theta \leqslant 2\pi$.

(b) State the period of the function.

(c) Give the coordinates of points of intersection with the θ-axis.

12 Using the graph of $y = \sin\theta°$, show that:

(a) $\sin(90 - \theta)° = \sin(90 + \theta)°$

(b) $\sin(180 - \theta)° = -\sin(180 + \theta)°$.

13 Using the graph of $y = \cos\theta°$, show that:

(a) $\cos\theta° = \cos(-\theta)°$

(b) $\cos(90 - \theta)° = -\cos(90 + \theta)°$

(c) $\cos(180 - \theta)° = \cos(180 + \theta)°$.

14 **(a)** Sketch the graph of $y = -\tan x°$, in the interval $-225 \leqslant x \leqslant 315$, showing the equations of any asymptotes.

(b) On the same set of axes, and for the same interval, sketch the graph of $y = \sin(x + 135)°$.

(c) Deduce the number of solutions, in the interval $-225 \leqslant x \leqslant 315$, of the equation $\tan x° + \sin(x + 135)° = 0$.

15 The diagram shows a sketch of part of the graph of $y = f(x)$. The point $R(3\pi, r)$ is a maximum point and the curve crosses the x-axis at $P(p, 0)$ and $Q(q, 0)$, where $q > p > 0$. Given that $f(x) = -\cos x$:

(a) state the value of r,

(b) write down, in terms of π, the value of:
(i) p **(ii)** q.

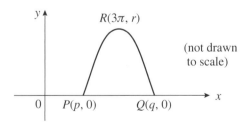

Given instead that $f(x) = \sin kx$, $k > 0$, write down:

(c) the smallest possible value of k,

(d) in terms of π, the value of:
(i) p **(ii)** q.

Test yourself	What to review
	If your answer is incorrect
1 In the interval $0 < x < 360$, write down the coordinates of any minimum points for the functions: **(a)** $\sin(x - 30)°$ **(b)** $2 + \cos x°$ **(c)** $\sin 2x°$.	*Review Heinemann Book C2 pages 118–124 Revise for C2 page 51 Example 3*
2 Write down the periods of the following functions: **(a)** $\tan \frac{3}{4}x°$ **(b)** $\sin \dfrac{5\theta}{2}$.	*Review Heinemann Book C2 pages 118–119, page 123 Revise for C2 page 51 Example 4*
3 Write down the equations of the asymptotes for the curve with equation $y = \tan 2x°$, in the interval $0 \leqslant x \leqslant 360$.	*Review Heinemann Book C2 page 119 Revise for C2 page 51 Example 4*
4 State the transformation which maps: **(a)** the graph of $y = \tan x°$ onto that of $y = \tan(x + 50)°$, **(b)** the graph of $y = \cos x°$ onto that of $y = \cos(-x)°$.	*Review Heinemann Book C2 pages 121–124 Revise for C2 page 51 Example 4, page 52 Worked exam style question 1*
5 The diagram shows part of the curve with equation $y = \sin 3\theta$. 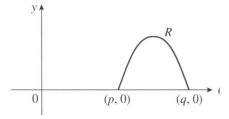 Given that $0 < p < \pi$, and that R is a maximum point: **(a)** find the value of p, **(b)** deduce the value of q, **(c)** write down the coordinates of R.	*Review Heinemann Book C2 page 123 Revise for C2 page 52 Worked exam style question 2*

Test yourself answers

1 (a) $(300, -1)$ **(b)** $(180, 1)$ **(c)** $(135, -1), (315, -1)$ **2 (a)** 240 **(b)** $\dfrac{4\pi}{5}$ **3** $x = 45, x = 135, x = 225, x = 315$

4 (a) Horizontal translation by -50 **(b)** Reflection in the y-axis (or horizontal stretch with scale factor -1)   **5 (a)** $\dfrac{\pi}{3}$ **(b)** π **(c)** $\left(\dfrac{5\pi}{6}, 1\right)$

Differentiation

9

Key points to remember

1 For an increasing function f(x) in the interval (a, b),
f$'(x) > 0$ in the interval a $\leqslant x \leqslant$ b.

2 For a decreasing function f(x) in the interval (a, b),
f$'(x) < 0$ in the interval a $\leqslant x \leqslant$ b.

3 The points on the graph of $y =$ f(x) where f(x) stops increasing and begins to decrease are called maximum points.

4 The points on the graph of $y =$ f(x) where f(x) stops decreasing and begins to increase are called minimum points.

5 A point of inflexion is a point where the gradient is at a maximum or minimum value in the neighbourhood of the point.

6 A stationary point is a point of zero gradient. It may be a maximum, a minimum or a point of inflexion.

7 To find the coordinates of a stationary point, find $\dfrac{dy}{dx}$, i.e. f$'(x)$, and solve the equation f$'(x) = 0$ to find the value, or values, of x and then substitute into $y =$ f(x) to find the corresponding values of y.

8 The stationary value of a function is the value of y at the stationary point. You can sometmes use this to find the range of a function.

9 You may determine the nature of a stationary point by using the second derivative.

If $\dfrac{dy}{dx} = 0$ and $\dfrac{d^2y}{dx^2} > 0$, the point is a minimum point.

If $\dfrac{dy}{dx} = 0$ and $\dfrac{d^2y}{dx^2} < 0$, the point is a maximum point.

If $\dfrac{dy}{dx} = 0$ and $\dfrac{d^2y}{dx^2} = 0$, the point is either a maximum or a minimum point or a point of inflexion.

> In this case you need to use the tabular method **10** and consider the gradient on either side of the stationary point.

If $\dfrac{dy}{dx} = 0$ and $\dfrac{d^2y}{dx^2} = 0$, but $\dfrac{d^3y}{dx^3} \neq 0$, then the point is a point of inflexion.

10 You can also find out whether a stationary point is a maximum, a minimum or a point of inflexion by drawing a table and calculating the gradient on either side of the point to establish the shape of the curve.

11 In problems where you need to find the maximum or minimum value of a variable y, first establish a formula for y in terms of x, then differentiate and put the derived function equal to zero to find x and then y.

Example 1

Show that the function $f(x) = x^3 + 6x^2 + 15x + 18$, $x \in \mathbb{R}$ is an increasing function.

$f(x) = x^3 + 6x^2 + 15x + 18$

$f'(x) = 3x^2 + 12x + 15$ ← First differentiate to obtain the gradient function

$f'(x) = 3(x^2 + 4x + 5)$

$f'(x) = 3[(x + 2)^2 + 1]$

As $(x + 2)^2 \geqslant 0$ for all real x

$3[(x + 2)^2 + 1] > 0$

∴ $f'(x)$ is an increasing function ← Using **1**: $f'(x) > 0$. As this is true **for all values** of x the curve **always** has a positive gradient

Example 2

Find the values of x for which the function $f(x) = 3 + 24x + 3x^2 - x^3$ is a decreasing function.

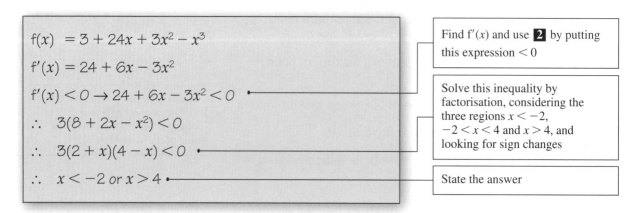

$f(x) = 3 + 24x + 3x^2 - x^3$

$f'(x) = 24 + 6x - 3x^2$

Find $f'(x)$ and use **2** by putting this expression < 0

$f'(x) < 0 \rightarrow 24 + 6x - 3x^2 < 0$

∴ $3(8 + 2x - x^2) < 0$

Solve this inequality by factorisation, considering the three regions $x < -2$, $-2 < x < 4$ and $x > 4$, and looking for sign changes

∴ $3(2 + x)(4 - x) < 0$

∴ $x < -2$ or $x > 4$ ← State the answer

Example 3

Find the least value of $x^2 - 5x + 3$. State the range of the function $f(x) = x^2 - 5x + 3$.

Let $\quad y = x^2 - 5x + 3$

Then $\quad \dfrac{dy}{dx} = 2x - 5$

Put $\quad \dfrac{dy}{dx} = 0$, then $x = 2.5$

$\therefore \quad y = 2.5^2 - 5 \times 2.5 + 3 = -3.25$

The least value of this quadratic function is -3.25 and the range is given by $f(x) \geqslant -3.25$

Using **7**: find the minimum point

There was only one turning point on this parabola and the question said that there was a least value, so you did not need to make a check.

The smallest point is the value of y at the stationary point

Using **8**: the range of the function is the set of values which y can take

Worked exam style question 1

Find the stationary points on the curve with equation $y = \dfrac{1}{x} + x + 3$.

Use the second derivative to establish whether the stationary points are maximum, minimum or points of inflexion.

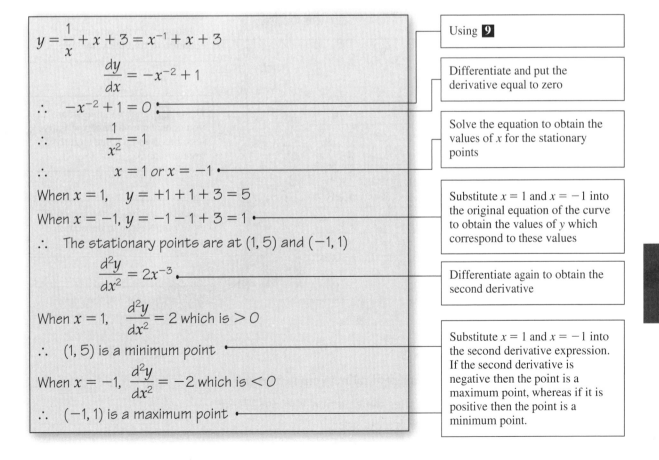

$y = \dfrac{1}{x} + x + 3 = x^{-1} + x + 3$

$\qquad \dfrac{dy}{dx} = -x^{-2} + 1$

$\therefore \quad -x^{-2} + 1 = 0$

$\therefore \qquad \dfrac{1}{x^2} = 1$

$\therefore \qquad x = 1 \text{ or } x = -1$

When $x = 1, \quad y = +1 + 1 + 3 = 5$

When $x = -1, \; y = -1 - 1 + 3 = 1$

$\therefore \quad$ The stationary points are at $(1, 5)$ and $(-1, 1)$

$\qquad \dfrac{d^2y}{dx^2} = 2x^{-3}$

When $x = 1, \quad \dfrac{d^2y}{dx^2} = 2$ which is > 0

$\therefore \quad (1, 5)$ is a minimum point

When $x = -1, \quad \dfrac{d^2y}{dx^2} = -2$ which is < 0

$\therefore \quad (-1, 1)$ is a maximum point

Using **9**

Differentiate and put the derivative equal to zero

Solve the equation to obtain the values of x for the stationary points

Substitute $x = 1$ and $x = -1$ into the original equation of the curve to obtain the values of y which correspond to these values

Differentiate again to obtain the second derivative

Substitute $x = 1$ and $x = -1$ into the second derivative expression. If the second derivative is negative then the point is a maximum point, whereas if it is positive then the point is a minimum point.

Worked exam style question 2

Find the coordinates of the stationary point on the curve with equation $y = x^3 - 9x^2 + 27x$ and establish whether it is a maximum or a minimum point or a point of inflexion.

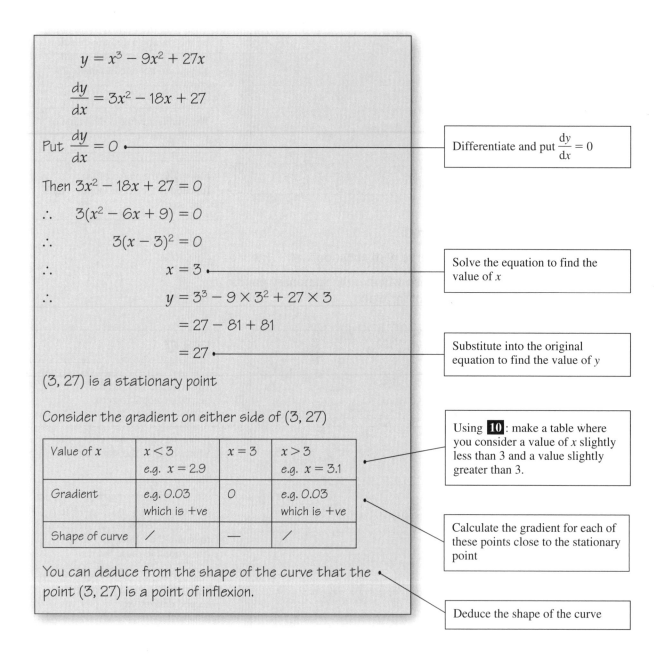

$y = x^3 - 9x^2 + 27x$

$\dfrac{dy}{dx} = 3x^2 - 18x + 27$

Put $\dfrac{dy}{dx} = 0$ — Differentiate and put $\dfrac{dy}{dx} = 0$

Then $3x^2 - 18x + 27 = 0$

$\therefore \quad 3(x^2 - 6x + 9) = 0$

$\therefore \qquad\qquad 3(x - 3)^2 = 0$

$\therefore \qquad\qquad\qquad x = 3$ — Solve the equation to find the value of x

$\therefore \qquad\qquad y = 3^3 - 9 \times 3^2 + 27 \times 3$

$\qquad\qquad\qquad = 27 - 81 + 81$

$\qquad\qquad\qquad = 27$ — Substitute into the original equation to find the value of y

$(3, 27)$ is a stationary point

Consider the gradient on either side of $(3, 27)$

Value of x	$x < 3$ e.g. $x = 2.9$	$x = 3$	$x > 3$ e.g. $x = 3.1$
Gradient	e.g. 0.03 which is +ve	0	e.g. 0.03 which is +ve
Shape of curve	/	—	/

Using **10** : make a table where you consider a value of x slightly less than 3 and a value slightly greater than 3.

Calculate the gradient for each of these points close to the stationary point

You can deduce from the shape of the curve that the point $(3, 27)$ is a point of inflexion.

Deduce the shape of the curve

Worked exam style question 3

The total surface area of a solid cylinder of radius r cm is $54\pi \,\text{cm}^2$.

(a) Show that its volume $V\,\text{cm}^3$ satisfies the equation $V = \pi r(27 - r^2)$.

(b) Find the maximum volume of such a cylinder.

(a) The total surface area is $2\pi r^2 + 2\pi rh$, where h is the height of the cylinder.

$\therefore \quad 2\pi r^2 + 2\pi rh = 54\pi$ •————

Divide both sides of the equation by 2π

$\therefore \quad$ Then $r^2 + rh = 27$

$\therefore \qquad\qquad rh = 27 - r^2 \qquad$ ① •————

$\qquad\qquad$ Volume $V = \pi r^2 h$

$\qquad\qquad\qquad = \pi r(rh)$

Use ① to give $V = \pi r(27 - r^2)$

Using **11** : establish V in terms of r only

The area of the circular top plus the circular base plus the curved surface

Make h or rh the subject of the formula

(b) $\qquad V = \pi r(27 - r^2)$

$\therefore \quad V = 27\pi r - \pi r^3$

$\therefore \quad \dfrac{dV}{dr} = 27\pi - 3\pi r^2$ •————

Expand the bracket and find $\dfrac{dV}{dr}$

$\qquad$ For maximum V, $\dfrac{dV}{dr} = 0$.

$\therefore \qquad\qquad 27\pi - 3\pi r^2 = 0$

$\therefore \qquad\qquad\qquad r^2 = 9$

$\therefore \qquad\qquad\qquad r = 3$ •————

As r is a length, $r > 0$. You do not need $r = -3$

$\therefore \quad$ Substitute to give $V = 54\pi$

This is a maximum value as $\dfrac{d^2V}{dr^2} = -6\pi r$, which is < 0 •————

Check the second derivative to ensure that V is a maximum

Thus the maximum volume of the cylinder is $54\pi \, \text{cm}^3$

Revision exercise 9

1 A curve C has equation $y = 2x^3 + 5x^2 - 4x + 3$.
Determine, by calculation, the coordinates of the stationary points of the curve C. Establish the nature of these stationary points.

2 A curve C has equation $y = \frac{1}{4}x^4 - 2x^3 + 4x^2 + 5$.
Determine, by calculation, the coordinates of the stationary points of the curve C. Establish the nature of these stationary points.

3 A curve C has equation $y = 60 - \dfrac{128}{x} - x^2$, $x > 0$.

Determine, by calculation, the coordinates of the maximum point of the curve C.

4 For the curve C with equation $y = x^4 - 8x^2 + 3$:

(a) find $\dfrac{dy}{dx}$,

(b) find the coordinates of each of the stationary points,

(c) determine the nature of each stationary point.

5 Given that $y = 125x^{\frac{5}{2}} - x^{\frac{1}{2}}$, $x > 0$, find the minimum value of y and show that it is a minimum.

6 $f(x) = 3x^5 - 5x^3$.

Find the values of x for which the function $f(x)$ is an increasing function.

7 The function f is defined by

$$f: x \to x + \frac{9}{x}, x \in \mathbb{R}, x \neq 0.$$

(a) Find $f'(x)$.

(b) Find the set of values of x for which f is an increasing function.

8 Given that $f(x) = x^3 - 9x^2 + 36x + 3$. Show that $f(x)$ is an increasing function.

9 $f(x) = \frac{1}{3}x^3 - x^2 - 63x + 6$.

Find the values of x for which the function $f(x)$ is a decreasing function.

10 The fixed point A has coordinates $(1, 3, 5)$ and the variable point P has coordinates $(t, t, 2t + 4)$.

(a) Show that $AP^2 = 6t^2 - 12t + 11$.

(b) Hence find the least distance between the points.

11 A square-based pyramid has slanting edges of length 6 cm. The volume of such a pyramid is $\frac{1}{3}x^2h$, where x is the length of the side of the base and h is the height. Show that this volume may be written in the form $24h - \frac{2}{3}h^3$. Prove that the greatest volume will occur when $h = 2\sqrt{3}$ cm.

12 An open rectangular box has a square base and has a volume of 500 cm³. It is made by sealing together along the edges of the square base and the four rectangular sides.

    (a) Show that the total length of the seal is $4x + \dfrac{2000}{x^2}$, where

        x cm is the length of the side of the square base.

    (b) Find the dimensions of the box if the total length of the seal is to be kept to a minimum.

13 A hollow cuboid container is constructed from the net shown in the figure. The dimensions of the container are x cm by y cm by z cm. The net is taken from a rectangular sheet of metal 25 cm by 20 cm as shown.

    (a) Show that $x = 12.5 - y$.

    (b) Establish a similar equation relating z and y.

    (c) Hence show that the volume of the cuboid is given by the relation $V = 2y^3 - 45y^2 + 250y$.

    (d) Find the maximum value of this volume, giving your answer to three significant figures.

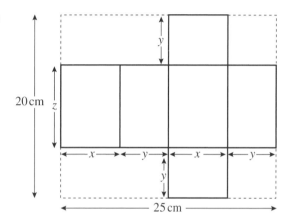

14 A container made from thin metal is in the shape of a right circular cylinder with height h cm and base radius r cm. The container has no lid. When full of water, the container holds 500 cm³ of water.

    (a) Show that the exterior surface area, A cm², of the container

        is given by $A = \pi r^2 + \dfrac{1000}{r}$.

    (b) Find the value of r for which A is a minimum.

    (c) Prove that this value of r gives a minimum value of A.

    (d) Calculate the minimum value of A, giving your answer to the nearest integer.   **E**

15 The figure shows a sketch of part of the curve C with equation $y = x^3 - 7x^2 + 15x + 3$, $x \geqslant 0$.

    The point P, on C, has x-coordinate 1 and the point Q is the minimum turning point of C.

    (a) Find $\dfrac{\mathrm{d}y}{\mathrm{d}x}$.

    (b) Find the coordinates of Q.

    (c) Show that PQ is parallel to the x-axis.   **E**

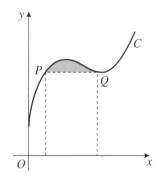

Test yourself	What to review

If your answer is incorrect

1 Find the values of x for which the following functions are increasing functions:

(a) $f(x) = 4x^3 + x^2 - 2x + 2$

(b) $f(x) = \dfrac{3}{x} - \dfrac{4}{x^3}$.

Review Heinemann Book C2 pages 129–130
Revise for C2 page 58
Example 1

2 Find the values of x for which the following functions are decreasing functions:

(a) $f(x) = (x - 3)(2x + 1)$

(b) $f(x) = \dfrac{3x + 1}{x^2}$.

Review Heinemann Book C2 pages 129–131
Revise for C2 page 58
Example 2

3 In each of the questions below, find the stationary points on the curve described by the equation, and determine whether these points are maximum, minimum or points of inflexion:

(a) $y = 3x^4 - 4x^3 - 12x^2 + 7$

(b) $y = x^3 - 6x^2 + 12x$

(c) $y = \dfrac{4}{x} - \dfrac{2}{\sqrt{x}}$

(d) $y = \dfrac{1 + 9x^2}{x}$.

Review Heinemann Book C2 pages 131–134
Revise for C2 page 59
Worked exam style question 1, page 60
Worked exam style question 2

4 Find the greatest possible value of $t^2(12 - t)$, for values of t between 0 and 12. Hence find the range of the function $f(t) = t^2(12 - t), 0 < t < 12$.

Review Heinemann Book C2 page 134
Revise for C2 page 59
Example 3

5 Given that x and y are positive variables such that $xy = 10$ and that $A = 5x + 2y$.

(a) Find an expression for A in terms of the single variable x.

(b) Find the least value of A.

Review Heinemann Book C2 pages 135–137
Revise for C2 page 60
Worked exam style question 3

Test yourself answers

1 (a) $x < -\frac{1}{2}, x > \frac{1}{3}$ **(b)** $-2 < x < 2$ **2 (a)** $x < \frac{5}{4}$ **(b)** $x < -\frac{2}{3}, x > 0$

3 (a) $(-1, 2)$ is a minimum point, $(0, 7)$ is a maximum point, $(2, -25)$ is a minimum point
(b) $(2, 8)$ is a point of inflexion   **(c)** $(16, -\frac{1}{4})$ is a minimum point   **(d)** $(\frac{1}{3}, 6)$ is a minimum point, $(-\frac{1}{3}, -6)$ is a maximum point

4 $256, 0 < f(t) < 256$ **5 (a)** $A = 5x + \dfrac{20}{x}$ **(b)** 20

Trigonometric identities and simple equations

10

Key points to remember

1 $\tan \theta = \dfrac{\sin \theta}{\cos \theta}$ (unless $\cos \theta = 0$, i.e. for $\theta = (2n + 1)90°$, when $\tan \theta$ is not defined)

> Reminder: For positive values of n, $(\sin \theta)^n$, $(\cos \theta)^n$ and $(\tan \theta)^n$ are written as $\sin^n \theta$, $\cos^n \theta$ and $\tan^n \theta$ respectively.

2 For all values of θ, $\sin^2 \theta + \cos^2 \theta = 1$.

3 A first solution of the equation $\sin \theta = k$, $-1 \leqslant k \leqslant 1$, is your calculator solution $\theta = \sin^{-1} k$.

A second solution is $180° - \sin^{-1} k$ (or $\pi - \sin^{-1} k$ if working in radians).

Other solutions are found by adding or subtracting multiples of $360°$ (or 2π radians).

4 A first solution of the equation $\cos \theta = k$, $-1 \leqslant k \leqslant 1$, is your calculator solution $\theta = \cos^{-1} k$.

A second solution is $360° - \cos^{-1} k$ (or $2\pi - \cos^{-1} k$ if working in radians).

Other solutions are found by adding or subtracting multiples of $360°$ (or 2π radians).

5 A first solution of the equation $\tan \theta = k \in \mathbb{R}$, is your calculator solution $\theta = \tan^{-1} k$.

A second solution is $180° + \tan^{-1} k$ (or $\pi + \tan^{-1} k$ if working in radians).

Other solutions are found by adding or subtracting multiples of $360°$ (or 2π radians).

Example 1

Simplify the following expressions:

(a) $\cos \theta \tan \theta$

(b) $3 \cos^2 2\theta + 3 \sin^2 2\theta$

(c) $\dfrac{2\sqrt{1 - \cos^2 3\theta}}{\cos 3\theta}$.

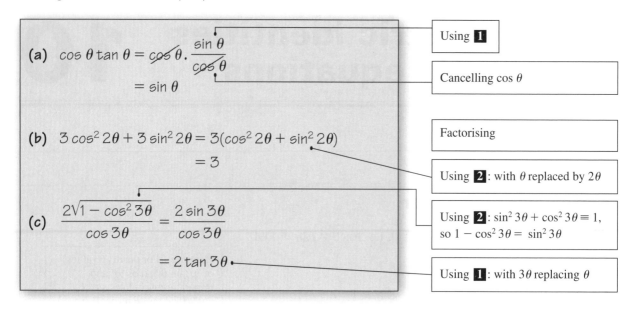

(a) $\cos\theta\tan\theta = \cancel{\cos}\theta.\dfrac{\sin\theta}{\cancel{\cos\theta}}$ Using **1**

$= \sin\theta$ Cancelling $\cos\theta$

(b) $3\cos^2 2\theta + 3\sin^2 2\theta = 3(\cos^2 2\theta + \sin^2 2\theta)$ Factorising

$= 3$ Using **2** : with θ replaced by 2θ

(c) $\dfrac{2\sqrt{1-\cos^2 3\theta}}{\cos 3\theta} = \dfrac{2\sin 3\theta}{\cos 3\theta}$ Using **2** : $\sin^2 3\theta + \cos^2 3\theta \equiv 1$, so $1 - \cos^2 3\theta = \sin^2 3\theta$

$= 2\tan 3\theta$ Using **1** : with 3θ replacing θ

Example 2

Prove that $\cos^4\theta + \sin^2\theta \equiv \sin^4\theta + \cos^2\theta$.

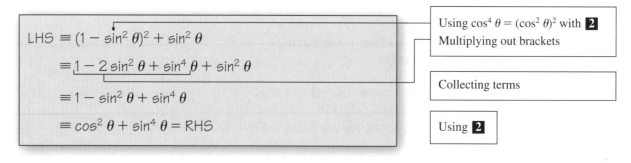

$\text{LHS} \equiv (1 - \sin^2\theta)^2 + \sin^2\theta$ Using $\cos^4\theta = (\cos^2\theta)^2$ with **2**. Multiplying out brackets

$\equiv 1 - 2\sin^2\theta + \sin^4\theta + \sin^2\theta$ Collecting terms

$\equiv 1 - \sin^2\theta + \sin^4\theta$

$\equiv \cos^2\theta + \sin^4\theta = \text{RHS}$ Using **2**

Example 3

Solve the following equations:
(a) $\cos\theta = \frac{1}{2}$, for $0 \le \theta \le 2\pi$ **(b)** $2\tan x = -3$, for $0 \le x \le 360°$.

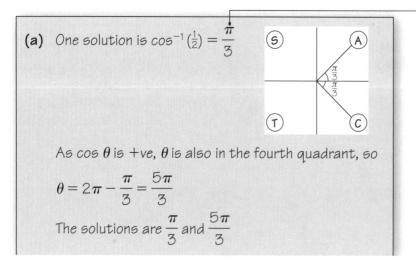

(a) One solution is $\cos^{-1}\left(\frac{1}{2}\right) = \dfrac{\pi}{3}$ This is your calculator value in terms of π

Remember that **A**, **S**, **T** and **C** give the quadrants for which **A**ll trig. functions, only **S**ine, only **T**angent and only **C**osine, are positive. (See page 48)

As $\cos\theta$ is +ve, θ is also in the fourth quadrant, so

$\theta = 2\pi - \dfrac{\pi}{3} = \dfrac{5\pi}{3}$

The solutions are $\dfrac{\pi}{3}$ and $\dfrac{5\pi}{3}$ Using **4** : solutions are $\cos^{-1}\left(\frac{1}{2}\right)$ and $2\pi - \cos^{-1}\left(\frac{1}{2}\right)$

(b) $\tan x = -1.5$

The calculator solution is $= -56.3°$ (3 s.f.)

The second solution is $x = 180° + (-56.3°)$

$\qquad = 123.7°$

The solutions are $123.7°$ and $303.7°$

(to the nearest $0.1°$)

> Dividing both sides by 2

> **Note:** This is not in the given interval; you will need to add $360°$

> Using **5**: $x = 180° + \tan^{-1}(-1.5)$

> In this case the solutions are $180° + \tan^{-1}(-1.5)$ and $360° + \tan^{-1}(-1.5)$

Example 4

Find all the values of θ in the interval $-180° \leqslant \theta \leqslant 180°$ for which:

(a) $\sin^2 \theta = \sin \theta$

(b) $2\cos^2 \theta + 3 \sin \theta = 0$.

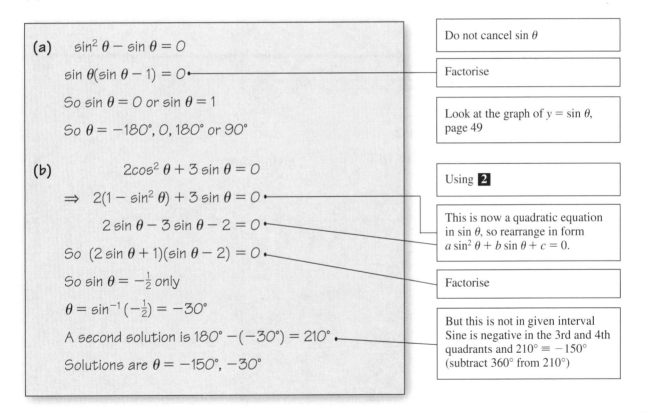

(a) $\sin^2 \theta - \sin \theta = 0$

$\sin \theta (\sin \theta - 1) = 0$

So $\sin \theta = 0$ or $\sin \theta = 1$

So $\theta = -180°, 0, 180°$ or $90°$

(b) $\qquad 2\cos^2 \theta + 3 \sin \theta = 0$

$\Rightarrow \quad 2(1 - \sin^2 \theta) + 3 \sin \theta = 0$

$\qquad 2\sin \theta - 3 \sin \theta - 2 = 0$

So $(2\sin \theta + 1)(\sin \theta - 2) = 0$

So $\sin \theta = -\frac{1}{2}$ only

$\theta = \sin^{-1}\left(-\frac{1}{2}\right) = -30°$

A second solution is $180° - (-30°) = 210°$

Solutions are $\theta = -150°, -30°$

> Do not cancel $\sin \theta$

> Factorise

> Look at the graph of $y = \sin \theta$, page 49

> Using **2**

> This is now a quadratic equation in $\sin \theta$, so rearrange in form $a \sin^2 \theta + b \sin \theta + c = 0$.

> Factorise

> But this is not in given interval Sine is negative in the 3rd and 4th quadrants and $210° \equiv -150°$ (subtract $360°$ from $210°$)

Worked exam style question 1

Given that $\cos \theta = \frac{3}{4}$ and that θ is reflex:

(a) find the exact value of $\sin \theta$,

(b) deduce the value of $\tan \theta$.

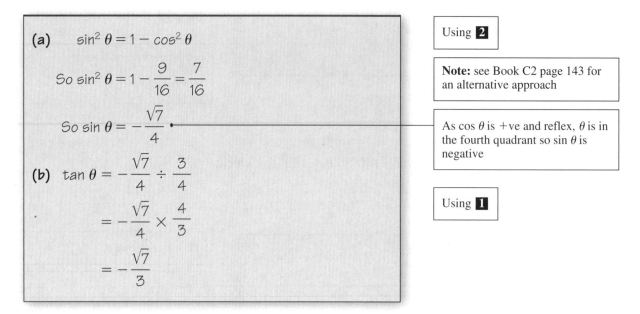

(a) $\sin^2 \theta = 1 - \cos^2 \theta$

So $\sin^2 \theta = 1 - \dfrac{9}{16} = \dfrac{7}{16}$

So $\sin \theta = -\dfrac{\sqrt{7}}{4}$.

(b) $\tan \theta = -\dfrac{\sqrt{7}}{4} \div \dfrac{3}{4}$

$ = -\dfrac{\sqrt{7}}{4} \times \dfrac{4}{3}$

$ = -\dfrac{\sqrt{7}}{3}$

Using **2**

Note: see Book C2 page 143 for an alternative approach

As $\cos \theta$ is +ve and reflex, θ is in the fourth quadrant so $\sin \theta$ is negative

Using **1**

Worked exam style question 2

Solve, in the interval $0 \leqslant \theta < \pi$, $\cos\left(3\theta + \dfrac{\pi}{8}\right) = \dfrac{\sqrt{2}}{2}$, giving your answers in terms of π.

Let $X = 3\theta + \dfrac{\pi}{8}$, so that $\cos X = \dfrac{\sqrt{2}}{2}$.

Since $0 \leqslant \quad \theta \quad < \pi$

$ 0 \leqslant \quad 3\theta \quad < 3\pi$

$ \dfrac{\pi}{8} \leqslant 3\theta + \dfrac{\pi}{8} < \dfrac{25\pi}{8}$

So the interval is $\dfrac{\pi}{8} \leqslant \quad X \quad < \dfrac{25\pi}{8}$

Calculator solution is $X = \cos^{-1}\dfrac{\sqrt{2}}{2} = \dfrac{\pi}{4}$

Other solutions in interval are $2\pi - \dfrac{\pi}{4}, 2\pi + \dfrac{\pi}{4}$

Using **4**: cosine is +ve in 1st and 4th quadrants

So $\quad X = \dfrac{\pi}{4}, \qquad \dfrac{7\pi}{4}, \qquad \dfrac{9\pi}{4}$

Since $X = 3\theta + \dfrac{\pi}{8}$, $\quad 3\theta = X - \dfrac{\pi}{8}$

So $\quad 3\theta = \dfrac{\pi}{4} - \dfrac{\pi}{8}, \quad \dfrac{7\pi}{4} - \dfrac{\pi}{8}, \quad \dfrac{9\pi}{4} - \dfrac{\pi}{8}$

$\qquad\quad = \dfrac{\pi}{8}, \qquad \dfrac{13\pi}{8}, \qquad \dfrac{17\pi}{8}$

So $\quad \theta = \dfrac{\pi}{24}, \qquad \dfrac{13\pi}{24}, \qquad \dfrac{17\pi}{24}$

Worked exam style question 3

(a) (i) Given that $5\sin x - 6\cos x = 0$, find the value of $\tan x$.
(ii) Hence find, to 1 decimal place, the solutions of
$5\sin x - 6\cos x = 0$, in the interval $0 \leqslant x < 360°$.
(b) (i) Show that $5\sin^2 y - 6\cos y = 0$ can be rewritten as $5\cos^2 y + 6\cos y - 5 = 0$.
(ii) Hence find, to 1 decimal place, the solutions of
$5\sin^2 y - 6\cos y = 0$, in the interval $0 \leqslant x < 360°$.

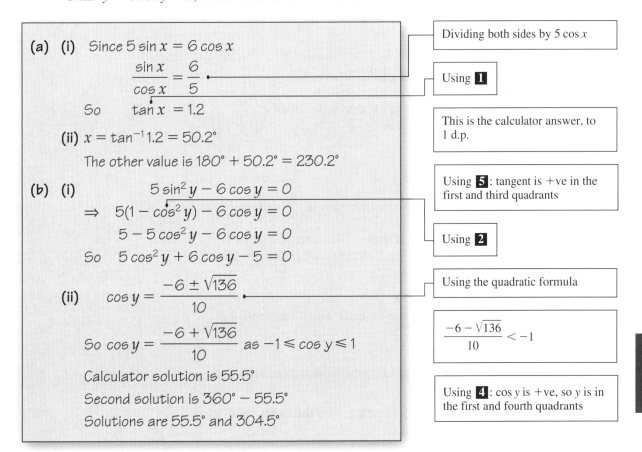

(a) (i) Since $5\sin x = 6\cos x$

$\qquad\qquad \dfrac{\sin x}{\cos x} = \dfrac{6}{5}$

So $\quad \tan x = 1.2$

(ii) $x = \tan^{-1} 1.2 = 50.2°$

The other value is $180° + 50.2° = 230.2°$

(b) (i) $\qquad\qquad 5\sin^2 y - 6\cos y = 0$

$\Rightarrow \quad 5(1 - \cos^2 y) - 6\cos y = 0$

$\qquad\quad 5 - 5\cos^2 y - 6\cos y = 0$

So $\quad 5\cos^2 y + 6\cos y - 5 = 0$

(ii) $\quad \cos y = \dfrac{-6 \pm \sqrt{136}}{10}$

So $\cos y = \dfrac{-6 + \sqrt{136}}{10}$ as $-1 \leqslant \cos y \leqslant 1$

Calculator solution is $55.5°$

Second solution is $360° - 55.5°$

Solutions are $55.5°$ and $304.5°$

Dividing both sides by $5\cos x$

Using **1**

This is the calculator answer, to 1 d.p.

Using **5**: tangent is +ve in the first and third quadrants

Using **2**

Using the quadratic formula

$\dfrac{-6 - \sqrt{136}}{10} < -1$

Using **4**: $\cos y$ is +ve, so y is in the first and fourth quadrants

Revision exercise 10

1 Simplify the following expressions:
   (a) $(\sin x + \cos x)^2 + (\sin x - \cos x)^2$
   (b) $\tan x \sqrt{1 - \sin^2 x}$.

2 Find the value(s) of $\tan x$, in each of the following equations:
   (a) $3 \cos x - 2 \sin x = 0$
   (b) $\dfrac{5 \sin x}{\cos x} = \dfrac{4 \cos x}{5 \sin x}$
   (c) $2(\sin x - \cos x) = 5 \cos x$.

3 Find, in the interval $0 \leqslant \theta < 360$, the solutions of:
   (a) $\cos \theta° = 0.65$
   (b) $\sin \theta° = -\dfrac{1}{\sqrt{2}}$
   (c) $2 \tan \theta° = 1$

giving your answers to 1 decimal place, if necessary.

4 Find, in the interval $-180° \leqslant \theta < 180°$, the solutions of:
   (a) $\cos \theta = 3 \sin \theta$
   (b) $\sin \theta = 2 \sin \theta \cos \theta$
   (c) $4 \cos^2 \theta = 3$

giving your answers to 1 decimal place, if necessary.

5 Given that θ is acute and $\sin \theta = k$, express in terms of k:
   (a) $\cos^2 \theta - \sin^2 \theta$
   (b) $\tan \theta \sin \theta$
   (c) $\cos^4 \theta$.

6 Given that $3 \sin x \cos y = 4 \sin y \cos x$, express $\tan y$ in terms of $\tan x$.

7 Over 24-hour periods the depth, d metres, of water in a canal is modelled by the formula $d = 6.5 + 2 \sin 15t°$, where t is the time in hours measured from 0800 h.
   (a) Find the value of d at mid-day.
   (b) Find, to the nearest minute, the times at which the depth of water is 5 m.

8 Given that $\sin x = \dfrac{5}{2\sqrt{7}}$, and that x is obtuse, show that $\tan x = -\dfrac{5\sqrt{3}}{3}$.

9 Given that $\cos x = -\frac{1}{3}$, and that x is reflex, find the exact value of:
   (a) $\sin x$ (b) $\tan x$.

10 Solve, in the interval $0 \leqslant \theta < 2\pi$, giving your answers in radians to 3 significant figures:

(a) $\sin 2\theta = 0.4$

(b) $\cos (\theta + 0.6^c) = -0.6$.

11 Solve, in the interval $0 \leqslant \theta < 360°$, giving your answers to 1 decimal place:

(a) $5 \sin \theta = \cos^2 \theta$

(b) $5 \cos \theta + 7 = 6 \sin^2 \theta$

(c) $5 \sin \theta = \tan \theta$.

12 The curve with equation $y = 2 - 6 \sin (x - 30)°$ crosses the y-axis at P. The nearest point to the origin at which the curve crosses the positive x-axis is Q.

(a) Write down the coordinates of P.

(b) Find the coordinates of Q.

13 (a) Given that $x = 2 \sin \theta$ and $y = \cos \theta$:

    (i) express $\tan \theta$ in terms of x and y,

    (ii) show that $x^2 + 4y^2 = 4$.

(b) Given that $x = \sin \theta + \cos \theta$ and $y = \sin \theta$, show that $x^2 - 2xy + 2y^2 - 1 = 0$.

14 Prove that:

(a) $2(1 - \cos \theta) - (1 - \cos \theta)^2 \equiv \sin^2 \theta$

(b) $(3 \sin \theta - 2 \cos \theta)^2 + (2 \sin \theta + 3 \cos \theta)^2 \equiv 13$.

15 Solve, in the interval $0 \leqslant \theta < 360°$, giving your answers to 1 decimal place:

(a) $\cos^2 2\theta = 0.16$

(b) $\tan (\theta + 40°) = \sin 20°$

(c) $3 \sin^2 \theta = 1 - \cos \theta$.

16 (a) Show that the equation $2 \cos \theta = 3 \tan \theta$ can be rewritten as $2 \sin^2 \theta + 3 \sin \theta - 2 = 0$.

(b) Hence solve $2 \cos \theta = 3 \tan \theta$, for $0 \leqslant \theta < 360°$, giving your answers to 1 decimal place.

17 (a) Factorise $2xy + 2x + y + 1$.

(b) Solve, in the interval $0 \leqslant \theta < 2\pi$, $2 \sin \theta \cos \theta + 2 \sin \theta + \cos \theta + 1 = 0$, giving your answers in terms of π.

Test yourself	What to review
	If your answer is incorrect
1 Simplify the following expressions: (a) $1 - \cos^2 2\theta$ (b) $3\cos^4 \theta + 3\sin^2 \theta \cos^2 \theta.$	*Review Heinemann Book C2* *page 142* *Revise for C2 page 65* *Example 1*
2 Show that $\tan^2 \theta - \dfrac{\sin^4 \theta}{\cos^2 \theta} \equiv \sin^2 \theta.$	*Review Heinemann Book C2* *page 142* *Revise for C2 page 66* *Example 2*
3 Given that $2\sin x° = -5\cos x°$: (a) write down the value of $\tan x°$, (b) solve, in the interval $0 \le x < 360$, $2\sin x° = -5\cos x°$, giving your answers to 1 decimal place.	*Review Heinemann Book C2* *pages 146–147* *Revise for C2 page 66* *Example 3, page 69* *Worked exam style question 3*
4 Solve, in the interval $-\pi \le \theta < \pi$, the equation $\cos\left(x + \dfrac{\pi}{6}\right) = \dfrac{1}{2}$, giving your answers in terms of π.	*Review Heinemann Book C2* *pages 149–150* *Revise for C2 page 68* *Worked exam style question 2*
5 Given that $\sin x = \frac{1}{4}$, and that x is obtuse, find the exact value of $\cos x$.	*Review Heinemann Book C2* *pages 143–144* *Revise for C2 page 67* *Worked exam style question 1*
6 Find all the values of θ, in the interval $0 \le \theta < 360°$, for which $2\cos^2 \theta + \sin \theta = 1$. Give your answers to 3 significant figures.	*Review Heinemann Book C2* *pages 151–152* *Revise for C2 page 69* *Worked exam style question 3*

Test yourself answers

6 $90°, 210°, 330°$

5 $-\dfrac{\sqrt{15}}{4}$

4 $-\dfrac{\pi}{2}, \dfrac{\pi}{6}$

3 (a) $-2\frac{1}{2}$ (b) $111.8, 291.8$

2 LHS $\equiv \tan^2 \theta - \tan^2 \theta \sin^2 \theta$
$\equiv \tan^2 \theta (1 - \sin^2 \theta)$
$\equiv \tan^2 \theta \cos^2 \theta$
$\equiv \sin^2 \theta =$ RHS

1 (a) $\sin^2 2\theta$ (b) $3\cos^2 \theta$

Integration

Key points to remember

1 $\displaystyle\int_a^b f'(x)\,dx = f(b) - f(a)$

2 The area between the curve with equation $y = f(x)$ and between the lines $x = a$, $x = b$ and the x-axis is

$$\text{Area} = \int_a^b f(x)\,dx$$

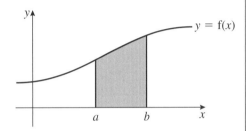

3 The area between a line (equation y_1) and a curve (equation y_2) is given by

$$\text{Area} = \int_a^b (y_1 - y_2)\,dx$$

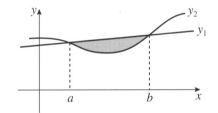

4 **Trapezium rule**

$$\int_a^b y\,dx \approx \tfrac{1}{2}h[y_0 + 2(y_1 + y_2 + \ldots + y_{n-1}) + y_n]$$

where $h = \dfrac{b-a}{n}$ and $y_i = f(a + ih)$.

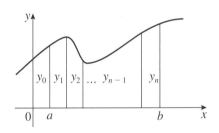

Example 1

Evaluate the following:

(a) $\displaystyle\int_2^3 \left(\frac{2}{x^2} - 3x^2 + 1\right) dx$

(b) $\displaystyle\int_1^4 (\sqrt{x} - 2)(1 - x)\,dx.$

(a) $\displaystyle\int_{2}^{3}\left(\frac{2}{x^2} - 3x^2 + 1\right) dx = \int_{2}^{3}(2x^{-2} - 3x^2 + 1)\, dx$

> Integrating in the usual way

$$= \left[\frac{2x^{-1}}{-1} - \frac{3x^3}{3} + x\right]_{2}^{3}$$

> Applying the limits in the usual way: substituting 3 first and then subtracting the result of substituting 2

$$= (-\tfrac{2}{3} - 27 + 3) - (-\tfrac{2}{2} - 8 + 2)$$

$$= -17\tfrac{2}{3}$$

> Definite integrals may give a negative answer

(b) $\displaystyle\int_{1}^{4}(\sqrt{x} - 2)(1 - x)\, dx = \int_{1}^{4}(x^{\frac{1}{2}} - 2 + 2x - x^{\frac{3}{2}})\, dx$

$$= \left[\frac{x^{\frac{3}{2}}}{\frac{3}{2}} - 2x + x^2 - \frac{x^{\frac{5}{2}}}{\frac{5}{2}}\right]_{1}^{4}$$

$$= \left[\tfrac{2}{3}x^{\frac{3}{2}} - 2x + x^2 - \tfrac{2}{5}x^{\frac{5}{2}}\right]_{1}^{4}$$

> Remember that $4^{\frac{3}{2}} = (\sqrt{4})^3 = 2^3$ and similarly $4^{\frac{5}{2}} = 2^5 = 32$

$$= (\tfrac{2}{3} \times 2^3 - 8 + 16 - \tfrac{2}{5} \times 2^5)$$

$$- (\tfrac{2}{3} - 2 + 1 - \tfrac{2}{5})$$

$$= \tfrac{16}{3} + 8 - \tfrac{64}{5} - \tfrac{2}{3} + 1 + \tfrac{2}{5}$$

$$= \tfrac{14}{3} + 9 - \tfrac{62}{5}$$

$$= 1\tfrac{4}{15} \text{ or } \tfrac{19}{15}$$

> You can use a graphical calculator to check your answers for definite integrals but in the C2 examination you should show all your working using the rules of integration or marks may be lost.

Example 2

(a) Find $\displaystyle\int 2x^2(x + 1)\, dx$.

(b) Hence find the area of the finite region between the curve with equation $y = 2x^2(x + 1)$ and the x-axis.

(a) $\displaystyle\int 2x^2(x + 1)\, dx = \int (2x^3 + 2x^2)\, dx$

> This is an indefinite integral like those met in Book C1

$$= \frac{2x^4}{4} + \frac{2x^3}{3} + c$$

(b) $y = 2x^2(x + 1)$

$0 = 2x^2(x + 1)$

So $x = 0$ or $x = -1$

So the curve crosses the x-axis at $(0, 0)$ and $(-1, 0)$

When $x = 0, y = 0$

$x \rightarrow \infty, y \rightarrow \infty$

$x \rightarrow -\infty, y \rightarrow -\infty$

A sketch of the curve will help you to see how to find the area

Put $y = 0$ and solve for x

$x = 0$ is a double root

Find the value of y when $x = 0$

Check what happens for large positive and negative values of x

Sketch the curve

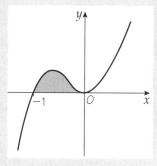

The area required is shaded and is given by

$$\text{Area} = \int_{-1}^{0} 2x^2(x + 1)\, dx$$

$$= \left[\frac{2x^4}{4} + \frac{2x^3}{3} \right]_{-1}^{0}$$

$$= (0) - \left(\frac{1}{2} - \frac{2}{3} \right)$$

$$= \frac{1}{6}$$

Using the integral from part **(a)**

Worked exam style question 1

The line with equation $y + x = 5$ cuts the curve with equation $y = 2x^2 - 9x + 11$ at the points A and B as shown in the diagram.

(a) Find the coordinates of the points A and B.

(b) Find the area of the shaded region between the curve and the line, as shown in the diagram.

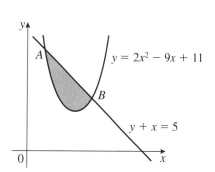

(a)
$$2x^2 - 9x + 11 = 5 - x$$
$$2x^2 - 8x + 6 = 0$$
So $\quad x^2 - 4x + 3 = 0$
$$(x - 3)(x - 1) = 0$$
So $\qquad\qquad x = 1 \text{ or } 3$
Therefore A is $(1, 4)$ and B is $(3, 2)$

The two equations are equal at A and B

Collect terms and simplify

Solve for x and use the equation of the line to find the y-coordinates, e.g. when $x = 1$, $y = 5 - 1 = 4$

(b) Shaded region $= \displaystyle\int_1^3 (\text{line} - \text{curve}) \, dx$

$$= \int_1^3 [5 - x - (2x^2 - 9x + 11)] \, dx$$

$$= \int_1^3 (-6 + 8x - 2x^2) \, dx$$

$$= \left[-6x + 4x^2 - \tfrac{2}{3}x^3 \right]_1^3$$

$$= (-18 + 36 - \tfrac{2}{3} \times 27) - (-6 + 4 - \tfrac{2}{3})$$

$$= 0 - -\tfrac{8}{3}$$

$$= \tfrac{8}{3} \text{ or } 2\tfrac{2}{3}$$

Worked exam style question 2

The line L has equation $y + x = 4$ and the curve C has equation $y = (x - 2)^2$. The line L meets C at the points $(0, 4)$ and A.
(a) Find the coordinates of A.
The region R is bounded by C, L and the positive x-axis.
(b) Find the area of R.

(a)

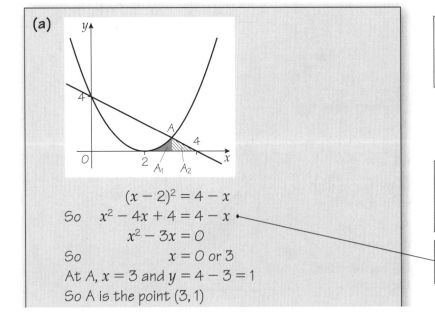

$$(x - 2)^2 = 4 - x$$
So $\quad x^2 - 4x + 4 = 4 - x$
$$x^2 - 3x = 0$$
So $\qquad\qquad x = 0 \text{ or } 3$
At A, $x = 3$ and $y = 4 - 3 = 1$
So A is the point $(3, 1)$

Both the equation of the curve and the equation of the line are quite simple so a sketch might help you see the area required.

Rearrange the equation of L to $y = 4 - x$ and then put line = curve to find where they cross.

Multiply out the brackets and collect together terms

From the sketch you can see that the required area $A_1 + A_2$

(b) $A_1 = \int_2^3 (x-2)^2\,dx$

This is the area to the left of $x = 3$ and under C

$= \int_2^3 (x^2 - 4x + 4)\,dx$

$= \left[\frac{x^3}{3} - 2x^2 + 4x\right]_2^3$

$= (9 - 18 + 12) - (\frac{8}{3} - 8 + 8)$

$= \frac{1}{3}$

$A_2 = \frac{1}{2}(4-3)(1)$

$= \frac{1}{2}$

A_2 is the area of the triangle

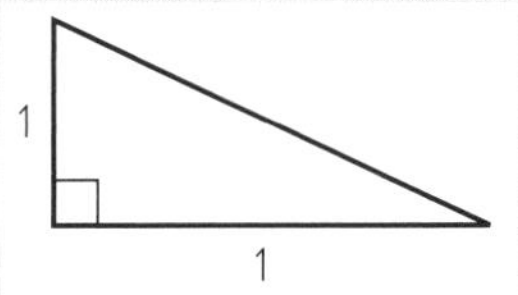

So the required area is $A_1 + A_2 = \frac{1}{3} + \frac{1}{2} = \frac{5}{6}$

Worked exam style question 3

A student uses the trapezium rule to evaluate I, where

$$I = \int_{0.5}^{1.5} \left(\frac{3}{x} + x^4\right) \mathrm{d}x.$$

(a) Complete the student's table, giving values to 2 decimal places where appropriate.

x	0.5	0.75	1	1.25	1.5
$\frac{3}{x} + x^4$	6.06	4.32			

(b) Use the trapezium rule, with all the values from your table, to calculate an estimate for the value of I.

(a) When $x = 1$, $\frac{3}{x} + x^4 = 3 + 1 = 4$

With the aid of a calculator, the other values are 4.8414 … and 7.0625

x	0.5	0.75	1	1.25	1.5
$\frac{3}{x} + x^4$	6.06	4.32	4	4.84	7.06

Complete the table

(b) $I \approx \frac{1}{2} \times \frac{1}{4}[6.06 + 2(4.32 + 4 + 4.84) + 7.06]$

$= \frac{1}{8}(39.44)$

$= 4.93$

Use the trapezium rule with $h = \frac{1}{4}$

Revision exercise 11

1 Evaluate $\int_1^2 \left(3x^2 - \dfrac{2}{x^3} + 6x\right) dx.$

2 Evaluate $\int_4^9 (10x^{\frac{3}{2}} - 6x^2 + 5)\, dx.$

3 Evaluate $\int_1^2 \left(3x^2 - \dfrac{2}{x^2}\right) dx.$

4 Evaluate $\int_0^4 (\sqrt{x} + 3)^2\, dx.$

5 Evaluate $\int_1^4 \left(\dfrac{\sqrt{x} + 3}{x^2}\right) dx.$

6 Evaluate $\int_1^4 \dfrac{x(5x - 3)}{\sqrt{x}}\, dx.$

7 Find the area of the finite region between the curve with equation $y = (x + 1)(4 - x)$ and the x-axis.

8 The finite region R is bounded by the line $x = 2$, the curve C and the positive x- and y-axes.
The equation of C is $y = x^3 - 3x^2 + 2x + 1$.
Find the area of R.

9 The diagram shows a sketch of the curve with equation

$$y = 2x + \dfrac{4}{x^3} - 3, \, x > 0.$$

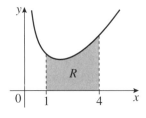

The region R is bounded by the curve, the x-axis and the lines $x = 1$ and $x = 4$.

(a) Find the area of R.

The region S is bounded by the curve with equation $y = 2x + \dfrac{4}{x^3}$, the x-axis and the lines $x = 1$ and $x = 4$.

(b) Write down the area of S.

10 The curve C has equation $y = (x - 3)^2$. The finite region R is bounded by C and the positive x- and y-axes.

(a) Find the area of R.

The point A lies on the negative x-axis and B is $(0, 9)$.
The triangle OAB has the same area as R.

(b) Find the coordinates of A.

11 The diagram shows a sketch of the curve C with equation $y = 6x - x^2$.

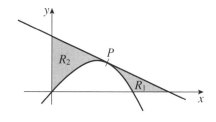

(a) Show that $P(4, 8)$ lies on C.

The tangent to C at P has equation $y + 2x = 16$.
The finite region R_1 lies between the tangent, C and the x-axis and the finite region R_2 lies between the tangent, C and the y-axis as shown in the diagram.

(b) Find the area of R_1.

(c) Find the area of R_2.

12 The diagram shows a sketch of the curve with equation $y = x^2 + k$. The point C is at $(2, 0)$ and the point D is at $(-2, 0)$. The points A and B lie on the curve and $ABCD$ is a rectangle. The curve divides the area of the rectangle in half. Find the value of k.

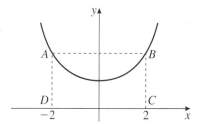

13 Use the trapezium rule with four strips to estimate

$$\int_0^2 \frac{1}{\sqrt{9 - x^2}} \, dx.$$

14 (a) Copy and complete the following table:

x	0	0.2	0.4	0.6	0.8	1.0
$y = 2^{-x}$	1		0.758	0.660		

(b) Use the results in the above table and the trapezium rule to estimate $\int_0^1 2^{-x} \, dx$.

15 The trapezium rule with the table below was used to estimate the area, A, between the curve with equation $y = (\sqrt{x} + 1)^2$, the lines $x = 0$ and $x = 4$ and the x-axis.

x	0	0.5	1	1.5	2	2.5	3	3.5	4
y	1	2.914	4	4.949	5.828	6.662	7.464	8.242	9

(a) Copy and complete the table, giving values to 3 decimal places where appropriate.

(b) Use the values from the table and the trapezium rule to find an estimate, to 2 decimal places, for A.

(c) Find the exact value of A.

(d) Calculate the percentage error in using the trapezium rule to estimate A.

Test yourself	What to review

If your answer is incorrect

1 Find $\displaystyle\int_{1}^{4}\left(6x + 3x^{\frac{1}{2}} - \frac{2}{x^2}\right)dx.$

Review Heinemann Book C2 pages 157–158
Revise for C2 page 73
Example 1

2

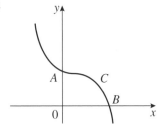

Review Heinemann Book C2 pages 159–161
Revise for C2 page 74
Example 2

The diagram shows a sketch of part of the curve C with equation $y = 4 + x^2 - x^3$.
The curve crosses the y-axis at the point A and the x-axis at B.

(a) Write down the coordinates of A.

(b) Show that B is $(2, 0)$.

(c) Find the area of the region bounded by OA, OB and C.

3

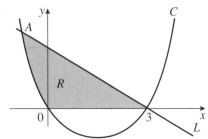

Review Heinemann Book C2 pages 164–167
Revise for C2 page 75
Worked exam style question 1, page 76
Worked exam style question 2

The diagram shows a sketch of the curve C with equation $y = x^2 - 3x$, and the line L with equation $y = 3 - x$.
The shaded region R is bounded by C, L and the x-axis.
The line L cuts C at A and $(3, 0)$.

(a) Show that A is $(-1, k)$ and find the value of k.

(b) Find the area of R.

4 Use the trapezium rule with four strips to estimate

$$\int_{-2}^{2} \sqrt{x^3 + x^2 + 4}\, dx.$$

Review Heinemann Book C2 pages 169–171
Revise for C2 page 77
Worked exam style question 3

Test yourself answers

1 57.5   2 (a) (0, 4)   (c) $\frac{20}{3}$ or $6\frac{2}{3}$ 3 (a) $k = 4$ (b) $-6\frac{1}{6}$ or $\frac{37}{6}$ 4 8.45

Examination style paper

1 (a) Find the first three terms, in ascending powers of x, of the binomial expansion of $(2 - 3x)^7$. **(4 marks)**

(b) Write down the last term of this expansion. **(1 mark)**

2 The first and third terms of a geometric series G are 100 and 36 respectively.

(a) Given that the common ratio, r, is positive, find the value of r. **(2 marks)**

(b) Find, to 1 decimal place, the sum of the first 15 terms of G. **(2 marks)**

(c) Find the exact value of the sum to infinity of G. **(2 marks)**

3 (a) Complete the following table:

x	0	0.5	1	1.5	2
$\dfrac{5}{x^2 + 1}$	5			1.538	

(2 marks)

(b) Use the trapezium rule, together with all the values from the above table, to estimate

$$\int_0^2 \frac{5}{x^2 + 1}\, dx.$$ **(4 marks)**

4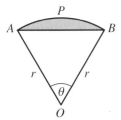

The diagram shows the cross-section through a traffic calming ramp. The curved surface, APB, of the ramp is in the shape of an arc of a circle centre O and radius r. The horizontal base, AB, of the ramp has length 100 cm and the highest point P of the ramp is 10 cm above AB.

(a) Show that $r = 130$ cm. **(3 marks)**

(b) Find the length of the arc APB, giving your answer to the nearest centimetre. **(4 marks)**

5

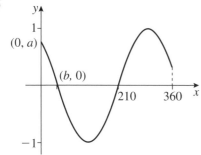

The diagram shows part of the graph with equation
$y = \cos(x + p)°$, where p is a constant. The graph crosses the
y-axis at $(0, a)$ and the x-axis at $(b, 0)$ and $(210, 0)$, where
a and b are constants.

(a) (i) Find the value of p and the value of b.   **(2 marks)**
     (ii) Write down the exact value of a.   **(1 mark)**

(b) Solve, for $0 < x < 360$, $\sin(x - 45)° = \frac{1}{2}$.   **(4 marks)**

6 (a) Solve $2^x = 19$, giving your answer to two decimal places.   **(2 marks)**

(b) Find the exact value of x that satisfies

$$\log_4(2 - 3x) - 2\log_4 x = 0.5.$$ **(6 marks)**

7 The points $P(0, 1)$, $Q(24, 1)$ and $R(24, 11)$ lie at the vertices
of a triangle T.

(a) Show that T is a right-angled triangle.   **(2 marks)**

The circle C passes through the points P, Q and R.

(b) State, giving a reason, which side of T is a diameter of C.   **(2 marks)**

(c) Find an equation for C.   **(6 marks)**

8

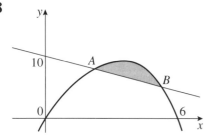

The line with equation $y + x = 10$ cuts the curve with equation
$y = x(6 - x)$ at the points A and B as shown in the diagram.

(a) Find the coordinates of the points A and B.   **(5 marks)**

(b) Find the area of the shaded region between the curve and
line, as shown in the diagram.   **(7 marks)**

9 A water trough is in the shape of an open cuboid with base dimensions x cm by y cm and height x cm.
The open surface has dimensions x cm by y cm.
Given that the capacity of the trough is 4500 cm^3:

(a) express y in terms of x, **(2 marks)**

(b) show that the external area, A cm^2, of the trough is given by
$$A = \frac{13\,500}{x} + 2x^2.$$ **(3 marks)**

The manufacturer needs to minimise the external area of the trough.

(c) Use calculus to find the value of x for which A is a minimum. **(5 marks)**

(d) Calculate the minimum value of A. **(2 marks)**

(e) Prove that this value of A is a minimum. **(2 marks)**

Answers to revision questions

Revision exercise 1

1 $\dfrac{2x+1}{x+4}$ **2** $(x+1)(3x^2-3x+2)$ **3** (b) $-5,-3,2$

4 $\frac{1}{2},5$ **5** (b) $-1,\frac{3}{2},2$ **6** $(x-4)(3x+2)^2$

7 (a) 7 (b) $8\frac{9}{16}$

8 (a) $(x-1)(2x-1)(3x+1)$ (b) $3x+1$

9 (a) $-13,6$ (b) $-3,\frac{1}{2},2$ **10** $f(q)=0$

11 (a) $p=-7,q=-14$ (b) $(x-2)(2x-1)(3x+4)$

12 (a) 3 (c) $x(2x-1)(x+2)$

13 (b) $(n+2)(n+1)(n+3)+3$

14 (a) -20 (b) 1

15 (a) $a=-2,b=5$ (b) $f(2)=0$

Revision exercise 2

1 (a) 62.8 (b) $48.6,131$ (c) 84.4 (d) 4.67

2 $1.24\,\text{cm}$ **3** 31.5 **4** $2610\,\text{m}^2$ **5** $4.07\,\text{cm}^2$

6 $x=53.2,y=16.9$ **7** 16.6 **8** $a=1,b=1$

9 1.80 **10** $19.6°$

11 (a) 7 **12** (b) 4

13 (a) $2.44\left(\dfrac{1+2\sqrt{10}}{3}\right)$ (b) $14.8°$

15 (b) $(x+\frac{1}{2})^2+36\frac{3}{4}$ (c) $6.06\,\text{cm}$ (3 s.f.), when $x=-\frac{1}{2}$

Revision exercise 3

1 (a) 4 (b) $1\frac{1}{2}$ (c) -1

2 (a) 8 (b) $\frac{3}{2}$

3 (a) $3+3m+n$ (b) $n-m-1$

3 (c) $m=4,n=5$ (d) $x=16,y=32$

4 2.23 **5** $0.77,1.39$ **6** 2.21 **7** $x=\frac{1}{4},y=8$

8 (a)

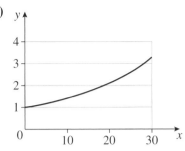

8 (b) £1184   **(c)** 17.7 years (3 s.f.)

9 (a) $4 + a$ **(b)** $4a - 1$ **(c)** $2\sqrt{2}$

10 (a) $2a - b$ **(b)** $2 + a + \frac{1}{2}b$ **(c)** $2a - b = 1, a + \frac{1}{2}b = -1$

10 (d) $-\frac{3}{2}$ **(e)** $x = 0.669, y = 0.089$

Revision exercise 4

1 (a) $(x - 1)^2 + (y - 4)^2 = 16$ **(b)**

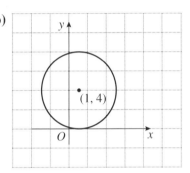

2 (a)

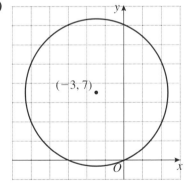

(b) $(0, 0), (-6, 0), (0, 14)$

3 (a) 8, 10   **(b)** 40

4 (a) $5\sqrt{2}$ **(b)** $(x - 2)^2 + (y + 1)^2 = 50$ **(c)** $y = x + 7$

5 (a) 8   **(b)** $\angle QPR = 90°$

6 (a) 80   **(b)** $(x + 2)^2 + (y - 4)^2 = 40$

7 (a) $3\sqrt{5}$ **(b)** $y = -\frac{1}{2}x + \frac{19}{2}$

8 (a) $(-2, -7), (6, 1)$ **(b)** $8\sqrt{2}$

9 (a)

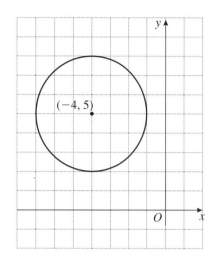

(b) $2\sqrt{10}$

10 (a) $a = 6, b = 8$ **(b)** $(x - 3)^2 + (y - 4)^2 = 25$

11 (a) $\angle PRQ = 90°$ **(b)** $(2, -1)$ **(c)** $(x - 2)^2 + (y + 1)^2 = 58$

12 (a) $(3, -3)$ **(b)** $x - 2y + 21 = 0$ **13 (a)** $(-5, -5), (3, 11)$

14 (a) 5 **(b)** $(-1, 0), (6, 7)$ **(c)** $\dfrac{\sqrt{2}}{2}$ **15 (a)** 52

15 (b) $y = \frac{1}{5}x + \frac{6}{5}$ **16 (a)** $(x - 5)^2 + (y - 13)^2 = 169$

16 (b) $5x + 12y = 62 = 0$ **17 (a)** $(4, 5)$

17 (b) $(x - 4)^2 + (y - 5)^2 = 25$ **(c)** 11.6 **18** $x^2 + (y - \frac{13}{2})^2 = \frac{25}{4}$

19 (a) $(3, -4), 10$ **(b)** $(9, 4)$ **20 (a)** $(4, 8), 17$ **(b)** $x = 13$

Revision exercise 5

1 (a) $81x^4 - 216x^3 + 216x^2 - 96x + 16$

1 (b) $32x^5 + 80x^4y + 80x^3y^2 + 40x^2y^3 + 10xy^4 + y^5$

2 $1 + 7x + 16x^2 + 8x^3 - 16x^4 - 16x^5$ **3** $a = \pm\sqrt{2}$

4 (a) $1 - 12x + 60x^2$ **(b)** $32 + 80x + 80x^2$

5 $A = 1024, B = 5120, C = 11\,520$

6 $1 + 18x + 135x^2 + 540x^3, x = 0.01, 1.19404$

7 (a) $A = 486, B = 540, C = 30$ **(b)** $x = \pm\sqrt{2}$

8 (a) $2016, 672$ **(b)** $k = 3$

Revision exercise 6

1 (a) (i) $120°$ **(ii)** $4.5°$ **(iii)** $140°$ **(iv)** $6.25°$

1 (b) (i) $\dfrac{\pi}{30}$ **(ii)** $\dfrac{3\pi}{40}$ **(iii)** $\dfrac{7\pi}{12}$ **(iv)** $\dfrac{7\pi}{3}$

2 (a) (i) $2.34\,\text{cm}$ **(ii)** $5\pi\,\text{cm}^2$ **(b) (i)** $3.042\,\text{cm}$ **(ii)** $43.75\pi\,\text{cm}^2$

3 **(a)** 2π cm   **(b)** 8π cm^2 **4** 38.7 cm^2 **5** 44.55 cm

6 0.64 radians   **7** 8 cm   **8** **(a)** 20 cm^2 **(b)** $67\frac{1}{2}\pi$ cm^2

9 $\frac{1}{2}r^2(1.2) = 28.8 \Rightarrow r = \sqrt{48} = \sqrt{16 \times 3} = 4\sqrt{3}$ **10** 6 cm^2

11 **(a)** $\dfrac{3\pi}{10}$ **(b)** **(i)** $(40 + 6\pi)$ cm   **(ii)** 60π cm^2 **(c)** 26.69... cm^2

12 Remember that $\sin(\pi - \theta) = \sin\theta$ **13** **(a)** 30.7 cm^2

13 **(b)** 21.9 cm   **14** **(b)** $\left(18 - \dfrac{9\pi}{2}\right)$ cm^2 **15** **(b)** 1070 m^2

16 **(b)** $P = 16.8$, $A = 14.4$ $(r = 6)$ **17** $r = 20$

18 **(c)** $r = 9$, $\theta = 2$

Revision exercise 7

1 **(a)** $r = 0.8$ **(b)** 0.83886   **(c)** 16.645   **2** **(b)** 4   **(c)** 0.00391

3 **(a)** 9, 27   **(b)** 12.9   **4** 0.5   **5** $r = 0.4$ or 0.6, $a = 30$ or 20

6 **(a)** $V = 16\,000\,(0.82)^a$ **(b)** £7234   **(c)** 11th year

7 **(a)** $p = 3$ **(b)** $r = 3$ **(c)** 59 048   **8** $r = 0.922$

9 **(b)** 306 minutes 41 seconds   **10** **(b)** 585

Revision exercise 8

1 **(a)** 1   **(b)** 0   **(c)** -1 **(d)** 1   **(e)** 0   **(f)** 1   **(g)** -1 **(h)** 1

1 **(i)** 0   **(j)** 0   **(k)** -1 **(l)** -1 **2** **(a)** $-\sin 80°$

2 **(b)** $-\cos 50°$ **(c)** $+\tan 5°$ **(d)** $-\sin\left(\dfrac{\pi}{6}\right)$ **3** 230, 310

4 **(a)** $-\frac{1}{2}$ **(b)** $+\dfrac{\sqrt{2}}{2}$ **(c)** $+\dfrac{\sqrt{3}}{2}$ **(d)** -1 **(e)** $-\frac{1}{2}$ **(f)** $+\dfrac{\sqrt{3}}{3}$

5 **(a)** 90   **(b)** 360   **(c)** 900   **(d)** 2π

6 **(a)** $x = -270$, $x = -90$, $x = 90$ **(b)** $x = 45$, $x = 225$, $x = 405$

6 **(c)** $x = -\dfrac{\pi}{6}$, $x = +\dfrac{\pi}{6}$, $x = +\dfrac{\pi}{2}$, $x = +\dfrac{5\pi}{6}$

7 **(a)**

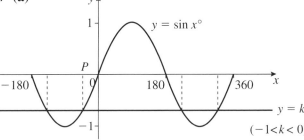

7 **(b)** $-180 - p$, $180 - p$, $360 + p$ **8** **(b), (c), (d), (e)**

9 (a) $y = \sin(x - 30)°$ **(b)** $y = 3 + \sin(x + 30)°$ **(c)** $y = \tan\dfrac{\theta}{3}$

10 (a) (i) 5 **(ii)** 270 **(b)** $y = 5\cos 2x°$

11 (a)

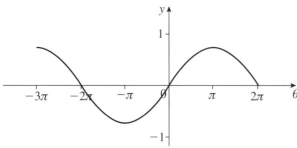

11 (b) 4π **(c)** $(-2\pi, 0), (0, 0), (2\pi, 0)$

12 (a)

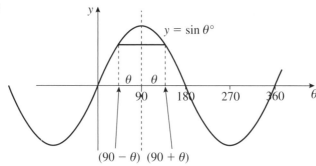

12 (b)

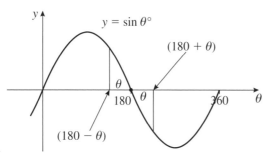

13 (a), (b)

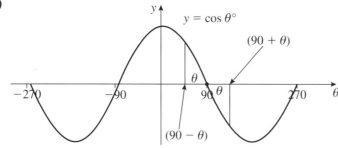

13 (c)

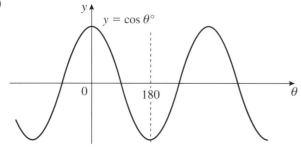

13 (a) There is symmetry in the line $\theta = 0$

13 (b) From the graph the y-coordinates at $(90 - \theta)°$ and $(90 + \theta)°$ are equal in magnitude and opposite in sign

13 (c) There is symmetry in the line $\theta = 180$, so at $(180 - \theta)$ and $(180 + \theta)$ y-coordinates are the same

14

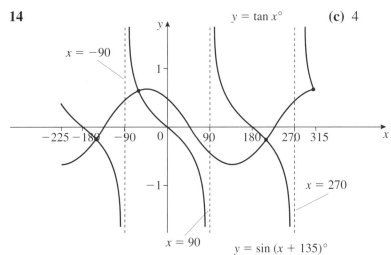

(c) 4

15 (a) 1　**(b) (i)** $\dfrac{5\pi}{2}$　**(ii)** $\dfrac{7\pi}{2}$　**(c)** $\frac{5}{6}$　**(d) (i)** $\dfrac{12\pi}{5}$　**(ii)** $\dfrac{18\pi}{5}$

Revision exercise 9

1 $(\frac{1}{3}, 2\frac{8}{27})$, minimum, $(-2, 15)$, maximum

2 $(0, 5)$ and $(4, 5)$ minima, $(2, 9)$ maximum　　**3** $(4, 12)$

4 (a) $4x^3 - 16x$　**(b)** $(0, 3), (2, -13), (-2, -13)$

4 (c) maximum, minimum, minimum　　**5** $(-\frac{4}{25})$

6 $x < -1, x > 1$　　**7 (a)** $1 - \dfrac{9}{x^2}$　**(b)** $x > 3, x < -3$

8 $f'(x) = 3[(x - 3)^2 + 3] > 0$　　**9** $-7 < x < 9$　　**10 (b)** $\sqrt{5}$

12 $10\,\text{cm} \times 10\,\text{cm} \times 5\,\text{cm}$　　**13 (b)** $z = 20 - 2y$

13 (c) $V = 410\,\text{cm}^3$　　**14 (b)** $r = 5.42$　**(d)** $A = 277$

15 (a) $3x^2 - 14x + 15$　**(b)** $(3, 12)$　**(c)** $y = 12$ at both P and Q

Revision exercise 10

1 (a) 2　**(b)** $\sin x$　　**2 (a)** $1\frac{1}{2}$　**(b)** $\pm\frac{2}{5}$　**(b)** $3\frac{1}{2}$

3 (a) 49.5, 310.5　**(b)** 225, 315　**(c)** 26.6, 206.6

4 (a) $-161.6, 18.4$　**(b)** $-180, -60, 0, 60$　**(c)** $\pm150, \pm30$

5 (a) $1 - 2k^2$　**(b)** $\dfrac{k^2}{\sqrt{(1 - k^2)}}$　**(c)** $1 - 2k^2 + k^4$　　**6** $\frac{3}{4}\tan x$

7 (a) 8.23 (3 s.f.)   **(b)** 04.46, 23.14   **9 (a)** $-\dfrac{2\sqrt{2}}{3}$ **(b)** $2\sqrt{2}$

10 (a) 0.206, 1.37, 3.35, 4.51   **(b)** 1.61, 3.47

11 (a) 11.1°, 168.9°   **(b)** 109.5°, 120°, 240°, 250.5°

11 (c) 0°, 78.5°, 180°, 281.5°   **12 (a)** (0, 5)   **(b)** (49.5, 0)

13 (a) (i) $\dfrac{x}{2y}$

13 (a) (ii) Using $\sin^2\theta + \cos^2\theta \equiv 1$, $\dfrac{x^2}{4} + y^2 = 1$ and then multiply through by 4

13 (b) $\cos\theta = x - y$ and $\sin\theta = y$, so $(x - y)^2 + y^2 = 1$ which leads to the result

14 (a) LHS $= (1 - \cos\theta)\{2 - (1 - \cos\theta)\}$
$\qquad = (1 - \cos\theta)(1 + \cos\theta)$
$\qquad = 1 - \cos^2\theta$
$\qquad = \sin^2\theta = $ RHS

14 (b) LHS $= (9\sin^2\theta - 12\sin\theta\cos\theta + 4\cos^2\theta)$
$\qquad\qquad + (4\sin^2\theta + 12\sin\theta\cos\theta + 9\cos^2\theta)$
$\qquad = 13\sin^2\theta + 13\cos^2\theta$
$\qquad = 13(\sin^2\theta + \cos^2\theta)$
$\qquad = 13 = $ RHS

15 (a) 33.2°, 56.8°, 123.2°, 146.8°, 213.2°, 236.8°, 303.2°, 326.8°

15 (b) 158.9°, 338.9°   **(c)** 0°, 131.8°, 228.2°

16 (a) $\qquad\qquad 2\cos\theta = 3\dfrac{\sin\theta}{\cos\theta}$

$\qquad \Rightarrow \qquad 2\cos^2\theta = 3\sin\theta$
So $2(1 - \sin^2\theta) = 3\sin\theta$
giving $2\sin^2\theta + 3\sin\theta - 2 = 0$

16 (b) 30°, 150°

17 (a) $(2x + 1)(y + 1)$ **(b)** $\pi, \dfrac{7\pi}{6}, \dfrac{11\pi}{6}$

Revision exercise 11

1 (a) 15.25   **2** -461 **3** 6   **4** 76   **5** 3.25   **6** 48

7 $20\frac{5}{6}$ **8** 2   **9 (a)** $7\frac{7}{8}$ **(b)** $16\frac{7}{8}$ **10 (a)** 9   **(b)** $A(-2, 0)$

11 (b) $6\frac{2}{3}$ **(c)** $21\frac{1}{3}$ **12** $\frac{4}{3}$ **13** 0.73

14 (a) 0.871, 0.574, 0.5   **(b)** 0.72

15 (a) 4, 7.464, 8.242   **(b)** 22.53   **(c)** $22\frac{2}{3}$ **(d)** 0.6%

Examination style paper

1 **(a)** $128 - 1344x + 6048x^2$ **(b)** $-2187x^7$ **2** **(a)** $\frac{3}{5}$

2 **(b)** 249.9 **(c)** 250 **3** **(a)** 4, 2.5 **(b)** 5.52 **4** **(b)** 103 cm

5 **(a)** **(i)** $p = 60, b = 30$ **(ii)** $\frac{1}{2}$ **(b)** 75, 195 **6** **(a)** 4.25

6 **(b)** $\frac{1}{2}$ **7** **(b)** *PR*, hypoteneuse of right-angled triangle

6 **(c)** $(x - 12)^2 + (y - 6)^2 = 13^2$ **8** **(a)** $A(2, 8), B(5, 5)$

8 **(b)** 4.5 **9** **(a)** $y = \dfrac{4500}{x^2}$ **(c)** 15 **(d)** 1350

9 **(e)** $\dfrac{\mathrm{d}^2 A}{\mathrm{d}x^2} = 4 + \dfrac{27000}{x^3}$